PRAISE FOR *100 UNI*

"The book is fantastic! *100 under $100* showcases human ingenuity solving seemingly intractable problems in elegant and sustainable ways. I felt my world expanding as I pored through the ideas that Betsy Teutsch has collected. What a valuable teaching tool—with writing so clear and accessible, students will be able to grasp complex problems and creative solutions; teachers will appreciate the endless ideas for service learning and meaningful involvement!"

—AMY J. COHEN, Director of Education, History Making Productions

"Beyond being an engaging and inspirational read, *100 Under $100* provides tangible ways of creating women's empowerment opportunities globally. In smart and accessible prose, Betsy Teutsch aptly underscores the powerful impact an organization can have if it takes into account the complex needs of the whole person. Teutsch's *100 Under $100* is an invaluable resource for those looking to dedicate their lives—or a few dollars—to the cause of women's empowerment worldwide."

—KENNEDY ODEDE, Founder and CEO, Shining Hope for Communities (SHOFCO)

"Betsy Teutsch has found some of the most brilliant yet simple tools that can be used in real-life settings to improve the lives of women challenged by poverty. While the writing is accessible to the lay reader and the examples selected are aimed at specific problems, the scope is sufficiently broad that even the most tech-oriented and development-savvy folks will find something new. The photos brilliantly illustrate the many faces of women who benefit from using these tools, and help make this book a pleasure to read."

—SUSAN HOLCK, MD, former Director of Reproductive Health, World Health Organization

"Women do the majority of the work in many communities of the Global South. Betsy Teutsch's book of one hundred tools is a smart and effective approach to building local capacity. Grouped into eleven sections ranging from health to subsistence farming to legal practices, the choice of tools is as insightful as it is practical, as interesting as it is important. Those new to this subject will find a heartening introduction. Those with more experience are unlikely to be aware of the full range of tools on offer."

—THOMAS JACOBSON, Director, Master of Science in Globalization and Development Communication program at Temple University

"Global health is the most rapidly expanding and exciting area of public health. Teutsch's book is a fascinating toolbox that will enthuse and empower public health students and professionals. This book effectively translates the usual theory-based curricular content to a virtual DIY learning lab. The range of doable and affordable tools for energy, sanitation, nutrition, agriculture, health, finance, and law will stimulate many to act positively to alleviate global poverty and gender-based inequalities."

—FRANK FRANKLIN, MD, MPH, PHD, Professor Emeritus of Public Health, University of Alabama at Birmingham

"*100 Under $100* is a formidable accomplishment. It should be mandated reading for people in the NGO and aid worlds; it is so full of interesting information valuable to everyone who cares about conditions in the developing world. Teutsch's writing is clear and succinct, and the organization is very reader-friendly. I especially like the book's identifiers—positives in green and negatives in red—as well as the action-oriented notes at the bottom of each entry. And the photos enliven it all!"

—BRIGHID BLAKE, African Sisters Education Collaborative /Sisters Leadership Development Initiative

"Betsy Teutsch's book *100 Under $100*—solutions to issues for women and girls in the developing world that work and that are relatively cheap—is an incredible read! You'll be amazed and inspired."

—DONNA SHAVER, Dining For Women – Vancouver, WA

100 Under $100

One Hundred Tools for Empowering Global Women

▲ photo © Sebastian Rich

Betsy Teutsch

SHE WRITES PRESS

Published 2015
Printed in China
ISBN: 978-1-63152-934-4
Library of Congress Control Number: 2014955652

Cover and interior design by Tabitha Lahr

COVER PHOTOGRAPHY CREDITS
Front cover: © Tanzeel Ur Rehman/Cover Asia Press, Courtesy of Photoshare. In rural Rajasthan, India, Vijaylaxmi Sharma, 24, escaped her own child marriage ten years ago and now helps save other young girls from this tragic cultural tradition
Front cover flap: A Zambian bike recipient, Ethel takes her cousin with her to school. © Brooke Slezak/World Bicycle Relief
Back cover (from left to right): Kangaroo caring mom with her twins, skin-to-skin © UN Foundation/Talia Frenkel; A wondrous inflatable LED light arrives in Haiti. © LuminAID; Vananh Le demonstrates plastic thatch roofing, piloted in Ecuador. © David Saiaa/Reuse Everything Institute; Rauha Heita, Bicycle Empowerment Network's lead bike mechanic, Okalongo, Namibia. © Michael Linke
Back cover flap author photo: © Margaret Shapiro

For information, address:
She Writes Press
1563 Solano Ave #546
Berkeley, CA 94707

Dedicated to all who engage
in *Tikkun Olam* — Repairing the World

Expanding the flow of justice
Fueling compassion
Generating love
Building peace

CONTENTS

◀ photo (facing page): Lakshmi Venkata, a widow who inherited her husband's land, supports herself and two sons, Andhra Pradesh, India. © Deborah Espinosa

Sector 3 — Lights, Cell Phone Charging, Energy! 30

Sector 4 — WASH = Water and Sanitation Hygiene 44

Sector 5 — Domestic Technology 64

Sector 6 — Subsistence Farming 74

Sector 7 — Construction 96

PREFACE AND ACKNOWLEDGMENTS

A challenge by Anglican Bishop Desmond Tutu hangs on my studio wall:

> "Do your little bit of good where you are; it's those little bits of good put together that overwhelm the world."
>
> —*Desmond Tutu*

I have been a professional calligrapher and artist for forty years, privileged to live in economic comfort and security in North America. Over the past six years, I have embarked on a new path, immersed in the extensive research that has culminated in this book. You may be wondering how I came to write about ways to help women across the globe climb up out of poverty.

Much as I admired idealistic peers who headed off to the Peace Corps back in the day, I was never so bold. I work from a home art studio and have long focused my volunteer time and organizational energy in my local community. For some time, though, I knew I wanted to be more engaged in global issues.

The emerging Internet provided access to many new arenas.

Pursuing new interests, I served as director of communications at GreenMicrofinance.org. Learning how poverty alleviation could be accomplished through eco-smart technologies bowled me over. Seeking to support women's engagement in designing and disseminating affordable, appropriate technology to improve their lives, I explored the overlap of the tech and women's empowerment sectors.

My first step was to learn about what was already being accomplished. I collected my material on Pinterest, then a new social media site for posting image-based collections. Users "pin" photos virtually, along with mini-commentaries, creating topical bulletin boards. Pinning images of women utilizing effective solutions proved to be an excellent and fun method of organizing content; I had no clear plan of what to do with the information. But I couldn't believe how much was out there, and how each initiative was a gateway to yet more exciting endeavors, all doing transformative work.

After weeks of exciting discoveries and avid pinning, I sat back, admiring these dazzling visuals: take-charge, inspiring women installing water treatment systems; building schools; and inventing ingenious, affordable solutions to huge problems. They were changing the trajectories of many lives. Though I have a tendency to be a bit of a Pollyanna, there was obviously a whole lot of genuinely great news to report.

It became crystal clear to me: **These women's stories and accomplishments deserve a wider audience**. In my mind's eye, I could see a beautiful book highlighting their efforts and solutions. The result is the volume now in your hands, a multiyear project that has thrilled me from beginning to end, and given me an opportunity to do my bit of good, where I am. I trust that readers will be inspired, too.

This is a book of stories showcasing—but not ranking—solutions to problems. Spanning eleven sectors, it reflects poverty's multifaceted nature, with no single pathway out, and follows the sectors that make up the world of global development. My goal is to raise awareness of the wide range of successful efforts to help women achieve healthier, more productive lives. Stacey Edgar, in her book *Global Girlfriend,* beautifully addresses privileged people putting forth opinions on how to alleviate poverty:

> ". . . The fact is I don't have any better answers than the women have already found for themselves. I am just blessed with more access to opportunity."
>
> —*Stacey Edgar*

As an artist, I was especially drawn to images of women wielding tools with strength and authority. Too often, our impressions of impoverished women are formed by pictures of passive female victims, lined up following disasters, awaiting aid. Disasters are a small subset of all the challenges faced by globally impoverished girls and women. The bigger story shared in this book is how global girls' and women's hard work—like testing water (check out tool **#38**), asserting their legal rights to inheritance (**#99**), organizing Village Savings and Loans (**#89**) or ninety-seven other types of efforts—are facilitating their own climbs out of extreme poverty, and bringing their families and communities with them.

Readers can learn about the photo research and collection process on p. 139.

An activist at heart, I've included suggestions for reader engagement, in addition to the implicit donation option. Each time I learned about a new great initiative, I wanted to jump in and volunteer; writing a book featuring 100-plus great solutions has allowed me to promote them all. I can't wait to hear from you, readers, as you chart your own service journeys.

Educators' enthusiasm about this material prompted me to suggest educational activities, included throughout the book. Watch for a Teacher's Guide in the future. Teachers and students: Please share your classroom initiatives and experiences at **www.100under100.org**.

As I complete *100 Under $100: One Hundred Tools for Global Women's Empowerment,* I am so thankful for all the help I have received.

I am greatly indebted to social media, facilitating my learning about and communicating directly with initiatives, humanitarian technologists, practitioners, development workers, NGO professionals, researchers, social entrepreneurs, journalists, photographers, and experts working hard, all over the globe—via Facebook, Flickr, Twitter, Pinterest, and Skype. What a gift of the 21st century!

Networking, both personal and digital, brought invaluable connections.

Elizabeth and Thomas Israel, the visionary creators of Greenmicrofinance.org, introduced me to appropriate, eco-smart technologies for the developing world. This book would not exist without their passion and pioneering hard work in this field.

Dana Raviv at Catchafire introduced me to the terrific initiatives at Technology Exchange Lab; our work was so congruent that we partnered on this project.

tel Where their logo appears in the book, it indicates that readers can learn more about a tool at TEL's extensive database.

Miranda Spencer connected me with She Writes Press, a publisher of quality books that empowers authors to become social entrepreneurs. Brooke Warner, the publisher, immediately saw the synergy between She Writes Press and the book's subject matter, wisely and patiently shepherding this inexperienced author through the

◀ photo (facing page): Leafy greens thrive on urine fertilizer at the Chisungu School's garden, in Epworth, Zimbabwe. © Peter Morgan/SuSanA

publication process. Designer Tabitha Lahr took complex page elements and worked her magic, producing an elegant design that does justice to the magnificent photos and guides readers through compact, detailed material.

I am immensely grateful to each and every photographer who granted a photo permission. Many were shared on Flickr by people who never expected their photos to appear in a book. Thanks to Danielle Yorko, my photo research assistant, who tracked seemingly endless photo permissions.

Thanks to the expert readers who added so much to the final text. Dr. Susan Holck provided not only generous, extensive professional analysis but also unending moral support. Thanks also to experts Rob Goodier, Aaron Greenberg, Emily Kunen, Ruthie Lissy Rosenberg, Lauren Shaughnessy, Paul Scott, and Hansdeep Singh. Their input was invaluable, but any mistakes—of course—are my own.

Thanks to the many cheerleaders who supported me during this process: Dining For Women friends, cherished friends of many decades, my minyan—your enthusiasm has buoyed me. Feedback from respondents to the prototype improved the book immeasurably. Special thanks to Fran, Joyce, Pam, and Tara, who helped me keep body and soul together as I forged ahead. A more extensive list of thanks is posted at **www.100Under100.org**.

And to my family: Thank you! Nomi Teutsch has been a faithful believer in the value of *100 Under $100*; she perfectly nailed the reader engagement icon: YOU

My sister Sally Koslow, author extraordinaire, provided helpful advice and encouragement. Rebecca Rosen and Zachary Teutsch supported this project in numerous ways, including delivering a prize after the finish line. Johanna Resnick Rosen's enthusiasm was heartening.

My husband, David, consistently asked excellent questions, providing advice on everything from adverbs to indexing, never once falling asleep during dinners with a partner sharing endless details of humanitarian tech. Helpmate, indeed! I am very blessed.

Every effort has been made to provide accurate information, but the status of initiatives changes quickly. I apologize for errors or omissions. *100 Under $100* has no business relationship with any of the featured products, nor does Technology Exchange Lab.

There are, of course, too many great stories to tell them all. I invite feedback; new content and updates will be posted frequently at **www.100under100.org**.

—Philadelphia, PA
August 2014

Growing flowers with drip irrigation in Myanmar © Proximity Designs

INTRODUCTION

What is the Vocabulary for Discussing Global Poverty?

Though many terms are used for describing the world's poorest people, none are universally agreed upon. Rich people in low-income countries live well. Poor people in rich countries suffer deprivation, but usually have access to basics like plumbing and safe water. This book is about the lowest-income women living in the world's lowest-income countries—but we need shorthand for that. I most frequently use "low-income" or "impoverished." Other relevant terminology includes:

- **Low-resource** describes regions lacking power and sanitation infrastructures; people living there are disadvantaged by both personal poverty and the absence of available services.
- **Developing world** indicates low-income regions lacking infrastructures, used in contrast with the industrialized world. (Many criticize this term, since it implies countries advance through set stages from lower to higher; in reality, development is much less linear, and the present state of the industrialized world is far from ideal.)
- **Global South** is a general term for the large swath of low- and middle-income regions where the vast majority of the world's poorest live: Africa, Asia, and Latin America.
- **Base of the Pyramid** is the largest, poorest global socio-economic group—3 billion people living on less than $2.50 (US) per day, both in rural and urban slum settings.
- **Third World**, a geopolitical term from the Cold War era, still refers to infrastructure-poor locales, though it is no longer used by development professionals.

What Do We Mean By Poverty Alleviation?

Standards of living in the 21st century are rising. Poverty is enormously challenging to measure, but all agree that hundreds of millions of people are moving up out of extreme poverty. However, there are hundreds of millions more who still live with great deprivation. Some argue that the poor will always be with us. In relative terms, that is true; there will always be those at the bottom of any economic distribution. But those who live with the fewest resources can make progress improving and stabilizing their lives when given access to affordable solutions.

Affluent people, far removed from those who live on $1 or $2 a day, tend to glaze over at statistics like "2.6 billion people lack sanitation" or "3 billion people cook directly over smoky campfires." These are numbing numbers; they make people feel guilty and helpless. The premise of *100 Under $100* is that anyone can do something to help people help themselves. However modest the effort, it makes a difference to someone.

This book highlights 100-plus low-cost/high-impact tools and practices to help solve problems, which, when unaddressed, keep people impoverished. These tools generally save time and/or expense, or eliminate harms or poverty traps—behaviors that perpetuate poverty. At the simplest level, I see the goal not as poverty alleviation per se, but working toward:

- Eliminating human suffering from preventable illness and hunger.
- Stabilizing precarious existence, helping people access decent living conditions, sanitation, food security, and education.
- Expanding opportunity, giving hard-working people paths to improve their circumstances and derive better returns on their hard work.
- Combating the abuse and exploitation of the world's poor, including violence against women and other forms of unjust treatment of impoverished people.

New approaches to humanitarian technology, designed for the planet's poorest, emphasize **co-creation** (see more on this on p. 144). People are not defined only by their material poverty; they also have many assets and deep knowledge about how to accomplish tasks. Poverty is as much a deficit of opportunity as material deprivation.

This book shares inspiring stories about tools that help accomplish tasks more quickly and effectively, making use of improved, affordable designs and practices. Cynthia Koenig's Wello water wheel (**#36**) transports water in half the time with less physical strain, a beautiful illustration of Matthew Ridley's observation:

> "In one sense this is the story of humanity, creating time. Human progress has been about innovating in ways that save time so that there is more time for something else. Prosperity is time saved."
>
> —Matt Ridley, *The Rational Optimist*

Excited as I am about elegant problem-solvers like microinsurance (**#91**) or rainwater baskets (**#35**), it is important to acknowledge that systemic discrimination blocks women's progress out of poverty. Many incidents of horrific violence toward girls and women, widely covered in the media, occurred during the research period of this book, expanding awareness about the seriousness of this problem. To give readers a broader view of gender-based inequities, I have included legal tools—as vital as health and mechanical tools for reducing suffering, expanding opportunity, and stabilizing living circumstances.

Why Focus on Global Women in Development?

Women are powerful agents of development**.** Investing in low-income girls and women is not just a moral imperative; it is excellent policy for poverty alleviation. This is because of the so-called Girl Effect: educating girls postpones marriage and childbearing, while preparing young women for higher-skill jobs. It also promotes development, because when women's income rises they invest more of it in their families than men do.

Young, impoverished, illiterate girls are frequent targets for early marriage, forcibly acquired by husbands (**#95**), joining their households as de facto servants. They typically bear children soon thereafter. Their pregnancies are higher-risk, and their children have few resources to escape another cycle of poverty.

Contrast this with keeping girls in school, where they can gain knowledge and skills, enabling them to land higher-paying jobs and contribute financially to their families of origin. They marry later, and when they do they have smaller families and invest more in each offspring. A virtuous cycle replaces a poverty trap. Intergovernmental agencies, NGOs, and philanthropies are working to expand girls' education. Doing so creates a positive trajectory moving forward; investing in girls' education is generally credited with fueling Asia's economic boom, for example.

Microcredit (**#88**) pioneered in helping women—often unskilled and illiterate—to earn income outside their homes. Women typically spend a significant portion of that income improving their families' diets, housing, and health care, and paying school expenses. This is how women pull up not only themselves but also their families and communities from extreme poverty.

When women earn money, it can help redress the imbalance of power in their households, giving wives more say in decision making. This benefits everyone; more information yields better decisions. Often it is husbands' indifference to domestic technology that carries the day. The family therefore does not invest any of its limited income on tools that would free up women's time, improve family health, and raise household income. Education, outreach, word-of-mouth, and role models all help encourage and embolden women to demand the upgrades that will improve their daily lives.

Each sector's introduction includes specific gender challenges, along with strategies for transcending them, which benefit everyone.

Can Global Poverty AND Climate Change Be Tackled Together?

100 Under $100 focuses on poverty alleviation with a lens on women. However, readers will note many entries with the L&G icon, signifying that the featured product or practice provides local and global environmental benefits. While these eco-benefits are secondary, they are significant. Poverty alleviation can be accomplished AND we can improve our planet's health; people improving their living conditions need not break the global ecological bank. This is in part because it is neither feasible nor desirable to replicate the industrialized world's inefficient energy and sanitation infrastructures.

Energy poverty, the paucity of available fuel and electricity that is so constraining for daily lives, can be partially surmounted through distributed, renewable energy, featured in Sector 3. Redesigning hyper-efficient, affordable tools with lower-energy demands will provide people the opportunity to utilize mechanized solutions. Therefore:

- People can move out of poverty without necessarily expanding their carbon footprint. Some will actually enjoy quality-of-life upgrades while contracting their footprints by, for example, replacing open-fire cooking with improved cookstoves (**#50**) and swapping out kerosene lanterns for solar lights (**#27**). One lone solar panel can power multiple lights, cell phone charging, and a radio, TV, or tablet computer. This is a significant rung on the ladder out of extreme deprivation.
- Environmentally minded philanthropists can combine support of eco-sustainability with poverty alleviation, a major two-for-one. Dozens of examples of such dual-benefit initiatives are featured in these pages.

Many of the featured tools and practices can be utilized in the affluent world, putting wasteful resource consumption on a diet.

Frugally engineered designs, devices, and practices provide value to energy-poor customers. Why not encourage their use in high-resource areas as well?

- Low-carbon energy-sipping cooking techniques are highlighted in Sector 5. Solar cooking tools and thermal retention can be utilized anywhere. Why should people use more energy than necessary to cook?
- Why do we turn lights on during the day when bottle lights (**#76**) can accomplish the same thing with no energy at all?
- Embrace Infant Warmers (**#3**) save low-birth and premature babies' lives, consuming just 1 percent of the energy required by incubators, plus babies can be held rather than isolated. Why not integrate them into our high-tech health care system?
- Solar Ear's rechargeable hearing aid batteries (**#14**) not only radically slash outlays for hearing aid batteries, they also save the resources of manufacturing and trashing single-use batteries. Why don't we sell them in affluent markets?
- Improved sanitation initiatives utilizing waste to generate bio-fuel (**#33**, **#45**, and **#46**) are a smarter approach than flushing waste, using potable water for this task, and then treating and discharging it; this process consumes resources instead of capturing waste's assets. We should better utilize eco-sanitation designs in industrialized settings.

If we are serious about poverty alleviation and climate change mitigation, smallholder farmers could be paid for eco-service provision. This is the ethos behind carbon offsetting, but much more can be accomplished if the will and the funding were aligned. Subsistence agriculture, Sector 6, has the potential to absorb global carbon through agroforestry (**#69**), while improving the productivity of smallholder farms. Biochar (**#57**), essentially sequestered carbon, improves soil structure. Improved cookstoves lessen stress on forests, allowing them to serve as carbon sinks. Thinking about the ecological benefits of poverty alleviation tools broadens global problem-solving opportunities.

The tools highlighted in this book focus on improving the lives of the world's most impoverished women, but additional audiences value eco-smart tools and practices. These constituencies include: people striving for more eco-sustainable lifestyles, low-income folks seeking do-it-yourself solutions, homesteaders and off-gridders who generate their own power and manage self-contained sanitation systems, outdoor adventurers who need the basics, and permaculture enthusiasts.

A comfortable life can be had anywhere by growing one's food in a keyhole garden (**#65**) constructed of bottle bricks (**#74**), and fertilized by urine (**#62**). Solar panels provide electricity (**#26**), but daytime light is bottled (**#76**). Rainwater is harvested (**#35**) and heated in a Solvatten (**#41**). Pedal a Giradora (**#56**) to do laundry, and you've exercised; toss the greywater right into the keyhole's compost. Cook with a solar box (**#51**) or parabolic stove (**#52**), augmented by a Wonderbag (**#53**). Add chickens (**#71**) and bees (**#70**); relax at night with drinks cooled in your ChotuKool (**#55**).

Individual and household efforts like these add up, shrinking carbon footprints without sacrificing quality of life. And great ideas for accomplishing this are coming from people all across the globe.

What are the Basics of Humanitarian Tech Design?

Humanitarian tech programs, teaching how to design products appropriate for low-resource settings, have burgeoned over the last decade. Stanford University's famous "Design for Extreme Affordability" two-semester course includes students from many disciplines and has been an incubator of innovative products, including the featured Embrace Infant Warmer (**#3**). MIT's hub, D-Lab, has likewise generated many initiatives, including the featured Zimba Water Chlorinator (**#37**) and biochar briquettes (**#33**). Many other universities offer classes and workshops, often including travel to the field to test products under real-world conditions.

The mantra of these efforts is "innovate, and then iterate, iterate, iterate"—tweaking prototypes to incorporate user feedback, and continually improving the final product. The most basic rule is that the products must work and be affordable; humanitarian technologists focus relentlessly on creating the simplest possible solutions.

Designs must be frugally engineered—stripped down, sturdy enough for rugged conditions and either manually powered or highly energy-efficient. The fewer the parts, the higher the likelihood of success; if products break with no one to repair them, or no replacement parts are to be had, the initiative fails. Solvatten water treatment boxes (**#41**) have only one part that needs replacing, a filter. Folded layers of an old sari can be used if no filter is available. One Dollar Eye Glasses (**#15**) are just lenses and wire frames, fabricated on the spot with a machine requiring no electricity.

The best designs are co-created with end users. There is increasing awareness that products donated to low-income individuals or communities without user input frequently do not perform as hoped. By working directly with local users, designers can explore an item's function and cultural context, assuring that the design will be more likely to gain acceptance. Greenway Grameen improved cookstove (**#50**) founders did extensive early-stage consulting with their target users; the resulting stoves have been extremely popular.

Products constructed with local materials, like Potters for Peace's ceramic water filters (**#42**), are the most sustainable in the long run. They have the added advantage of providing livelihoods and developing local knowledge and capacity.

Aesthetics and appeal need to be considered. Lucky Iron Fish™ (**#54**) had a proven solution for combating iron deficiency: an iron lump added to pots while cooking. End users were unenthusiastic. Molding it in an attractive good-luck fish shape has increased sales and acceptance significantly.

What are Social Businesses?

A great deal of attention has been focused on sales and distribution of useful products to low-income customers: affordable goods that people purchase, rather than receive for free, typically have a faster, wider uptake. Business, obviously, has deeper pockets for investing in product development and advertising than the nonprofit sector. Competition between companies lowers prices and improves products.

C. K. Prahalad's 2004 book *Fortune at the Bottom of the Pyramid: Eradicating Poverty Through Profits* highlighted companies succeeding with a low-income customer base, identifying methods for harnessing business to deliver much-needed products. The book was written before the runaway success of cell phones in the developing world. Cell phone penetration—more people own cell phones than toilets—has upended thinking about what low-income customers can and will purchase.

Social entrepreneurs, merging social good with commercial distribution, are growing in number and influence. Social businesses launched by idealistic entrepreneurs strive to do well by doing good. Like strictly commercial ventures, they focus on their bottom lines and customers' needs and behaviors, but they also seek to improve customers' lives; they generally set eco-sustainability goals as well. Some strive for a triple bottom line: people, profits, and planet.

Some social businesses are hybrid organizations, with both a for-profit sales division and a nonprofit entity to which supporters of their mission can donate, subsidizing them while they strive to become economically sustainable. An example of this approach is Quetsol, a Guatemalan solar company (**#29**).

Humanitarian technologists are encouraged to think about marketing challenges during the design process to assure that new solutions will be reliable, affordable, and appealing.

What are the Barriers to Distributing Proven, Affordable Solutions?

While the expansion of social businesses is a promising trend for delivering affordable, well-designed products that improve quality of life, it is not a panacea. Particularly when it comes to health-related interventions, Sectors 1 and 2, distribution is too often fragmented and poorly managed. Low-income families often cannot afford medications along with clinic or hospital visits.

Low-resource regions' public health services are frequently underfunded, understaffed, and understocked. This is one of the reasons that children, as well as adults, die from preventable diseases, despite the availability of inexpensive (or free, in some jurisdictions) drugs such as three featured in this book: Oral Rehydration Therapy for diarrhea (**#11**), artemisinin therapy for malaria (**#9**), and Misoprostol for post-partum hemorrhage (**#22**).

NGOs and intergovernmental organizations work tirelessly to help improve health delivery, but cannot possibly provide enough care without strong in-country commitments to improving health care quality and access.

These truths have been highlighted for me by following the progress of Project Muso, founded by Drs. Ari Johnson and Jessica Beckerman, friends of my son and daughter-in-law. Project Muso, in partnership with Tostan, Partners in Health, and the Malian Ministry of Health, concentrates on eliminating preventable deaths in ultra-poor communities. Based in Yirimadjo, Mali, a country with one of the world's highest child-death rates, Johnson applies intense focus to the challenge of global health distribution. He observes:

> "We have a forty-year delivery backlog in the world's poorest communities, and the impact of these incredible tools that could save millions of lives annually will depend entirely on whether we are able to build effective, equitable delivery systems."
>
> —*Dr. Ari Johnson*

Project Muso launched an ethnographic study interviewing mothers about their approach to malaria treatment. While medication in Mali is free, treatment requires a trip to a clinic and a $1.50 payment—the equivalent of a family's entire daily income.

> "We discovered that mothers weren't accessing free life-saving antimalarial medications, not primarily because of ignorance but rather as the result of a sophisticated and heartbreaking analysis they did every day to try to keep their children alive, as they faced financial, geographic, and infrastructure barriers to getting their children the resources they need to survive.
>
> As we researched why so many children continue to die in Mali of curable diseases like malaria, pneumonia, and diarrhea, we found that the great challenge was delivery. Imagine you buy a computer. You install your favorite applications, and press the power button, but nothing happens. Then you realize why—there is no operating system effectively installed. Just so, we have found that life-saving technologies are only as good as their delivery system. The tools only work if they are embedded in a system that enables mothers to access them and use them at the right time."
>
> —*Dr. Ari Johnson*
> *(via e-mail with the author)*

Project Muso's approach to lowering child deaths—incorporating community health outreach workers, deputizing locals to refer ill children, and including income-generating projects—has yielded impressive drops in child deaths. A study outlining their approach and results is posted at PlosOne.org.

The supply chain problem bedevils business as well—how can great products make their way to remote locations? One promising development in medication delivery is the microfranchise door-to-door agent (**#92**); these vendors sell affordable—but lifesaving—medications as part of their for-profit product offerings. Because microfranchises are connected to commercial supply chains, their vendors are often better stocked with basic medications than clinics or hospitals. With modest training, cumulative on-the-ground experience, and one-on-one relationships with their customers, they are effective providers of basic health information and tools to end users. And yes, they even make house calls!

How Can Tools Be Combined for More Impact?

Common sense suggests that multiple tools are better than one. Some of the 100 tools pair logically, multiplying their impact, such as these examples:

- ✔ Improved cookstoves (**#50**) can use fuel briquettes formatted to fit them (**#32**, **#33**).
- ✔ Rainwater baskets (**#35**) collect the water for micro-drip irrigation (**#64**).
- ✔ In some locales, birth registrations (**#94**) are paired with vaccinations (**#4**).
- ✔ Schools are deputized to provide lunch (**#8**), calories that help students thrive. Schools also distribute micronutrient supplements (**#7**) and deworming meds (**#10**).
- ✔ Nutrient-rich SASHA sweet potato vines (**#59**) are given out as premiums at prenatal visits (**#19**), boosting the health of pregnant mothers, families' older children, and fathers, too.

Nonprofits can integrate services combining practices and tools from different sectors. Two such entities are profiled below, one in an urban slum setting, and the other serving rural subsistence farmers.

As a founding board member of Kibera's Shining Hope for Communities, SHOFCO.org, headed by the dynamic Kennedy Odede and Jessica Posner, I observed their simultaneous moves forward on many fronts to create a high-functioning oasis and anchor in Kibera, Africa's largest slum. This informal settlement is adjacent to, but not incorporated into, Nairobi. SHOFCO uses a pioneering cross-sector community development model in Kibera, an area completely lacking basic infrastructure.

SHOFCO has based its activities around the high-quality, free Kibera School for Girls that educates low-income, high-achieving girls (**#81**), and a health clinic providing general care and dispensing antiretroviral medicines for HIV/AIDS patients (**#13**). The clinic also provides family planning services (**#18**) and cervical cancer screening (**#17**). They stock a supply of ready-to-use therapeutic food (**#12**) for the most at-risk children in the community.

Graduation day at the Kibera School for Girls, Kenya. @ *Candace Hope, Transit Authority Figures*

The school serves snacks and lunch, (**#8**) and tooth brushing is included in the school day (**#6**). SHOFCO runs a microfinance program for community members, (**#88**) as well as savings circles (**#89**). They teach computer literacy classes in their cyber café (**#86**), which is open for daily use. Their eco-sanitation bio-center accommodates 300 users a day; the school's lunches are cooked with smoke-free, clean gas from the eco-toilets' biodigester (**#31**). They've just added a rooftop solar panel (**#26**) to power the library's lights. Sanitary napkins are distributed as needed (**#47**).

SHOFCO received funding to build improved latrines around the neighborhood (**#46**) and—quite a feat!—they facilitated running a water main from Nairobi, providing clean, treated water that can be transported home from the water tower on the school's mini-campus.

SHOFCO has famously pushed hard against young girls' rapists. Sadly, this crime is commonplace in Kibera. Holding the police more accountable and helping parents organize a neighborhood patrol and protests when known rapists are treated with impunity (**#96**) has helped improve the safety of women and girls.

The students of the Kibera School for Girls achieve at world-class rates, impressive for anywhere—even more remarkable for a school comprised of girls from a Global South slum's lowest socio-economic rung.

Why are the girls so successful? Teasing apart all the different impacts would be impossible. Small class size and sophisticated, interactive curricula are likely important. But nutritious food, clean water, sanitation, safety from sexual predators, increased respect for girls and women, antiretrovirals for infected patients, family planning, and microfinance enterprises for impoverished women all contribute to student success, community health, and poverty alleviation—as does charismatic leadership. SHOFCO's leaders would like to replicate its successful multi-sector model beyond Kibera.

Duplicating their success and financing it are large challenges.

The rural-based One Acre Fund, OneAcreFund.org, utilizes an integrated approach, working with more than 300,000 smallholder farmers. One Acre Fund provides inputs—high-quality seeds and fertilizer, chiefly—supported by its own microlending infrastructure (**#88**). Repayment schedules are flexible and aligned with harvest income; crop microinsurance (**#91**) is included in One Acre's offerings. Deliveries include tree seeds (**#69**), solar lamps with cell phone charging that can be sold as a service by enterprising farmers (**#27**), and triple bags for crop storage (**#67**). Participants receive extensive education and training in improved agro-methods and marketing strategies.

One Acre Fund does not provide direct health services, but works to strengthen local family-planning services in its farmers' communities (**#18**). One Acre Fund analyzes its metrics to determine if their bundled services help farmers escape poverty; increased yields, higher incomes, and excellent loan repayments are indeed the hoped-for results.

What Can I Do to Help?

100 Under $100 doesn't just report on inspiring solutions; each entry includes suggestions for YOU, the reader, to jump in and help. Many entries feature mission-driven nonprofits who would be thrilled to receive donations, of course. Quite a few list additional ideas on their websites for supporters to advocate and act on their behalf, including raising funds from others in creative ways.

The entries' featured suggestions range from simple actions that take only a few minutes, like purchasing fair trade products (**#93**) or tree planting (**#69**), all the way to starting new organizations or launching social enterprises (whole new careers). Readers will find student internship opportunities, as well as initiatives seeking *pro bono* volunteers to work both domestically and on-site in the developing world.

A list of volunteer opportunities for professionals culled from the 100 entries is on p. 143. Did you know flight attendants can serve as extra eyes to combat sex trafficking (**#97**)? Travel-related activities are included, too. How about taking your family to help build a bottle school in Guatemala (**#74**)? A list of travel-related activities is on p. 142.

Readers can connect directly with initiatives by "liking" their Facebook pages. You will then be in the loop, informed about their developments. Twitter handles are included, a great way to contact initiatives directly with ideas or questions. Even if organizations that grab your attention don't list specific volunteer opportunities, contacting them via Twitter or Facebook and asking, "What can I do to help?" would be deeply appreciated.

Opportunities for service, involvement, and engagement will be updated and posted at **www.100Under100.org**. Please share reports of your activities, and send pictures, too. Initiatives are invited to share new volunteer opportunities and wish lists.

100 TOOLS FOR EMPOWERING GLOBAL WOMEN: A GUIDE

What are the Criteria for Inclusion in *100 Under $100*?

This book's definition of "tool" is expansive. It includes health interventions, financial tools, and legal instruments. Directed to lay audiences, the entries are:

- ✔ **Replicable:** Many initiatives are excellent candidates for export to other regions; these opportunities are noted.
- ✔ **End user-focused:** Profiled tools are easy to understand and implement.
- ✔ **Proven:** Promising tools without a track record are mentioned, but not featured.

Why $100?

The $100 limit telegraphs the fact that small amounts can make serious poverty alleviation inroads. Here are 100 (and actually many more than that!) ways that women can improve their health and work their way out of extreme poverty, although it is important to note that for end users, $100 is an unimaginably large sum of money. Most of the entries are far less costly, and often pay for themselves in the long run.

Prices range from free, like SODIS (**#39**) and Urine Fertilizer (**#62**), to equipment slightly over $100 but shared by cooperatives, bringing the per-user cost well below $100.

Tools like Literacy (**#81**) are impossible to quantify in strictly dollar terms. Prices in some cases are the cost of materials, like Fishing Nets (**#73**). In other cases, prices are for delivered products. Products that are beyond what end users can pay upfront generally provide financing strategies or extend credit directly.

How Are the Items Organized?

The taxonomy of separating by sector can be tricky, since functions overlap. Where to put Plastic Bottle Solar Lights (**#76**), for example? Lights are covered under Energy, Sector 3, but these built-in lights provide daytime illumination via sunlight, using no energy at all, and are in Construction, Sector 7.

Working through a gender lens, I added a Domestic Technology sector focusing on women's work tools. This sector does not formally exist in the development and aid world. The lead item, Improved Cookstoves (**#50**), is usually a stand-alone sector, which overlaps with clean energy. Looking at things from a female perspective, subjects like cooking tools and laundry deserve more serious attention.

Horseback house call by Justina Munoz Gonzalez, vaccinating a baby living near the remote village of San Pablo, Nicaragua. © Adrian Brooks, Courtesy of Photoshare

ENTRY FEATURES AND SYMBOLS: A LEGEND

Checklists

For quick visual communication, note that:

- ✔ Indicates positives.
- ✗ Indicates negatives.
- • Indicates neutral information.

Challenges

When one's mission is promoting a particular tool, solution, or initiative—however effective it is—there is a natural reluctance to disclose negatives. It is, however, useful to understand barriers to success. Learning more about the problems of design and distribution is central to improving performance. Challenges are included to add information, not to critique approaches.

Icons

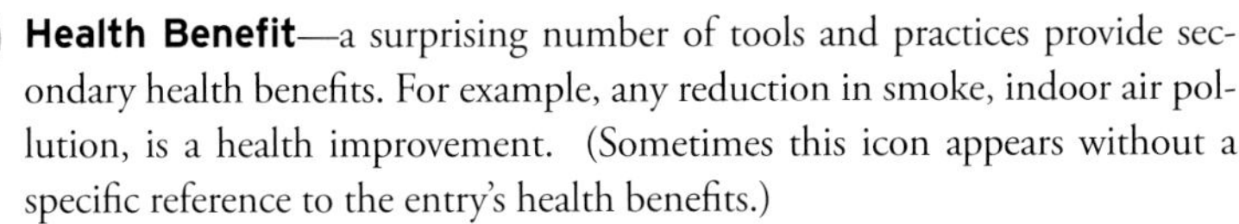

Health Benefit—a surprising number of tools and practices provide secondary health benefits. For example, any reduction in smoke, indoor air pollution, is a health improvement. (Sometimes this icon appears without a specific reference to the entry's health benefits.)

Do It Yourself—items that can be assembled from readily available, local materials.

Leapfrog Tech—skips over all earlier versions, straight to cutting-edge products, such as solar lamps (**#27**) and cell phones (**#84**).

Local and Global Eco-Benefits—indicates these tools or practices provide environmental benefits, although that is not their primary purpose. Read more about eco-smart tech on p. xiv.

Global South to Global North—tech that travels the reverse of the conventionally assumed direction. Read more about this phenomenon in Solar Plastic Bottle Lights (**#76**).

tel **Technology Exchange Lab**—indicates items posted at **www.technologyexchangelab.org**'s humanitarian technology data base. TEL curates information provided by developers, manufacturers, and recommenders, with regard to accuracy and relevance. Inclusion of a product or method in TEL's database does not imply endorsement.

Classroom ideas for teachers and students.

Suggested action items that you, the reader, can do. See more on pgs. 142-143.

Bolding

Bolding indicates featured tools and initiatives. Terms that are bolded when first introduced appear in the Glossary, p. 142. In the digital edition, bolding generally is hyperlinked.

Bolded numbers refer to **Tools #1 - #100.**

Website links for featured initiatives appear at the top of each page, along with Twitter handles (@xxxx). E-mails and Facebook pages can be found at their websites.

Sources

Direct sources are cited. When using general, public domain information, I conducted a "search engine test." If, when putting facts and general statistics into a search engine window, they popped up immediately at multiple locations, I did not cite a specific source.

Featured Bios

Most design is collaborative, but certain innovators stand out, inspiring individuals who dream up ingenious, remarkable solutions, and work tirelessly to implement them. Women are grossly underrepresented in STEM fields (Science, Technology, Engineering, and Math). When designers are women, I have highlighted them, honoring them and putting them forth as role models. (Hats off to the many men who contribute great solutions; they are in the book, too, of course.)

Sector 1 – An Overview

GENERAL HEALTH

Investment in health is the cornerstone of development. When health improves, families' incomes usually improve, too. Less time sick and fewer resources devoted to care and treatment means more opportunities for work and study. When incomes rise, people eat better and gain more access to education, further raising their health profile. Allocating resources to basic health care yields dividends for individuals, families, communities, and national economies.

These fifteen inspiring, low-tech, high-impact health tools potentially scale the formidable barriers of cost and distribution, as well as transcend the shortage of trained health professionals in low-resource settings. Utilizing longstanding, well-known interventions, along with cutting-edge innovations and entrepreneurial strategies, these entries get the job done. For example:

- ✔ Deworming (**#10**) uses schools for delivery, sidestepping weak health systems.
- ✔ ColaLife (**#11**) takes an underutilized, highly effective treatment, Oral Rehydration Therapy, packages it beautifully, and moves it from poorly stocked, underfunded clinics to for-profit kiosks, with excellent uptake results.
- ✔ Designing inexpensive eyeglasses (**#15**), educating people about their benefits, and selling them locally has created a robust social enterprise delivering better vision at prices affordable to low-income customers.

General health tools are gender-neutral, all things being equal. However, things are not equal. Sector 2 focuses on the health of girls and women specifically.

- ✗ Girls frequently receive less health care, from conception onward. Girl fetuses are more frequently aborted (illegally). Malnourishment causes a host of health maladies, and many families devote fewer resources, including food, to daughters. Unschooled girls are deprived of not just education but also access to school-based health centers' public health resources.
- ✗ Women bear the burden of caring for sick family members. Grandmothers, mothers, daughters, daughters-in-laws, sisters, and other female relatives care for parents, spouses, siblings, and children. This work is unpaid and demanding, especially if the women themselves are in poor health.
- ✗ Women often have little agency in family health care decisions, including the seeking of treatment, despite being primary caregivers.
- ✗ Ill health in pregnancy negatively impacts offspring. A malnourished mother is at risk for problem pregnancies with adverse outcomes for herself, her infant, and her other children if she dies. Investing in maternal health improves the health of both sons and daughters.

Health is not only promoted by health interventions. H: "Health promoting" icons show up in surprising entries throughout *100 Under $100.* Nearly a quarter of the 100 entries are explicitly health-related, but other interventions provide significant health benefits in addition to their main functions. For example:

- ✔ Solar Lights (**#27**) and Improved Cook Stoves (**#50**) decrease respiratory diseases caused by smoke inhalation, as well as lower the risk of burns.
- ✔ Rolling water rather than carrying it, or at least using the Pack H_2O with its more ergodynamic design, decreases stress on girls' and women's bodies (**#36**). Bikes (**#78**) and wheelbarrows (**#79**) offer similar secondary health benefits.
- ✔ Companion planting with nutritious legumes addresses micronutrient deficiencies (**#60**).
- ✔ Cell phones (and literacy) provide targeted health reminder texts (**#84**).
- ✔ A birth certificate provides access to government health benefits (**#94**).

There are many paths to healthier, more resilient lives!

1. BREASTFEEDING

The global campaign to promote breastfeeding stresses breast milk's ideal protective nutrition and highlights artificial formula's deadly risks in low-resource settings.

2. KANGAROO CARE

Kangaroo mother care—placing the baby skin-to-skin against the chest—raises survival rates of premature and low birth weight babies.

3. EMBRACE INFANT WARMER

Embrace Infant Warmers fight hypothermia in preterm and low birth weight babies for 1 percent of the cost of an incubator.

4. VACCINE DELIVERY

Due to weak vaccination delivery systems, many of the world's poorest children still die from preventable diseases. A coordinated global effort to raise vaccination rates is saving many lives.

5. UMBILICAL CORD TREATMENT

Newborns are vulnerable to infections. A single 25 **cent** application of CHX gel on the umbilical cord stump can save a baby's life.

6. TOOTH BRUSHING

Tooth brushing—a cheap, low-tech intervention—helps reverse the pandemic of children's tooth decay caused by over-consuming junk foods and sugared drinks.

7. MICRONUTRIENT-FORTIFIED FOOD AND SUPPLEMENTS

Fortifying staple foods with essential micronutrients, an inexpensive process, prevents blindness, mental retardation, and maternal deaths caused by micronutrient deficiencies.

8. SCHOOL LUNCH

Free school lunches improve student attendance and performance and are a proven incentive for educating daughters.

9. MALARIA PREVENTION AND TREATMENT

Insecticide-treated nets prevent malaria. Inexpensive, quick diagnosis and treatment is available.

10. DEWORMING

A vast majority of children in low-resource areas suffer from intestinal worms. Deworming capsules improve health and school attendance at a cost of 50 cents per child.

11. ORAL REHYDRATION SALTS + COLALIFE

Oral Rehydration Salts have saved millions of lives by preventing dehydration, which frequently causes death from diarrhea in malnourished people.

12. PLUMPY'NUT™: READY-TO-USE THERAPEUTIC FOOD

Peanut-based Ready-to-Use Therapeutic Food (RUTF) has revolutionized the care of acutely malnourished children, saving vastly more lives than was previously possible.

13. ANTIRETROVIRAL DRUGS: EXTENDING LIFE WITH HIV/AIDS

Though antiretrovirals extend the lives of HIV-positive patients, two-thirds of people who need them do not receive them.

14. SOLAR EAR: AN AFFORDABLE HEARING AID WITH RECHARGEABLE BATTERIES

90 percent of the 360 million hearing-impaired people in the world do not have access to hearing aids like the affordable and rechargeable Solar Ear, powered by solar-charged batteries.

15. VISION CORRECTION

Hundreds of millions of adults and children in the world lack access to affordable eyeglasses that can boost their literacy and their productivity.

A nurse counseling session, Cusco, Peru. ©Carmen Pfuyo Cahuantico/WABA

Breastfeeding

The global campaign to promote breastfeeding stresses breast milk's ideal protective nutrition and highlights artificial formula's deadly risks in low-resource settings.

waba.org.my • worldbreastfeedingweek.org • BabyMilkAction.org

1 What radically increases infant survival, for free? Sometimes the solution is hiding in plain sight: breastfeeding. It's available 24/7 without preparation, providing:

- ✔ Ideal baby food
- ✔ Critical warmth for tiny newborns, especially when nursing commences within an hour of birth
- ✔ Antibodies to fight diarrhea and pneumonia, major infant-killers
- ✔ Bonding and a sense of security
- ✔ Natural contraception; exclusive breastfeeding suppresses ovulation in 98 percent of mothers

> "There is no other single health intervention that has such a high impact for babies and mothers as breastfeeding and which costs so little for governments. Breastfeeding is a baby's 'first immunization' and the most effective and inexpensive lifesaver ever."
>
> —*Geeta Rao Gupta, UNICEF Deputy Executive Director*

Exclusive breastfeeding for the first six months of life could save hundreds of thousands of babies each year. Despite endorsement by medical authorities, governments, and legions of "lactivists," (proponents of lactation,) global breastfeeding rates hover at just 40 percent.

What happened to breastfeeding? Infant formula was introduced in the early 20th century; bottle-feeding quickly spread worldwide. Marketers touted its convenience, and it became fashionable and aspirational. It does not offer all of breastfeeding's benefits, but in high-resource regions with safe water, it presents no dangers.

Infant formula was aggressively marketed in low-income countries, pushing false claims about its benefits without mentioning its risks in low-resource settings. Mixing infant formula with unsafe water means babies are fed harmful bacteria while simultaneously being deprived of their mothers' infection-fighting antibodies, doubly dangerous. Babies' lives are endangered when:

- ✗ Unsafe water is added to formula powder, and the infants ingest pathogens.
- ✗ Bottles are washed with untreated water and transmit disease.
- ✗ Formula is diluted to stretch supplies, and babies become malnourished and more vulnerable to infection.
- ✗ Formula is tainted and babies are poisoned.

In the early 1970s, the nascent **World Association of Breastfeeding Action (WABA)**, headquartered in Malaysia, organized a global boycott of Nestlé, a multinational company infamous for unethical infant formula marketing in low-income countries. This lead to the landmark International Code for Marketing Breastmilk Substitutes adopted in 1981, which:

- ✔ Requires disclosure about the superiority of breastfeeding
- ✔ Bars public promotion of breast milk substitutes
- ✔ Prohibits health workers from pushing formula
- ✔ Bans free formula samples for pregnant women or new mothers

So far, eighty-four countries have passed these laws; fourteen more are in the process of doing so. The United States has not taken action on the code.

In order to increase breastfeeding's popularity, it's important to understand why mothers don't breastfeed. Successful breastfeeding requires education and support.

Obstacles include:

- Lack of advance preparation or ongoing coaching
- Pain and discomfort (especially when moms are lacking information on managing it)
- Stress when babies are slow to latch on
- Unsupportive medical advisers who encourage supplemental bottles, undermining milk supply
- Unaccommodating workplaces where it is difficult to express milk while away from the baby
- Lack of refrigeration to safely save expressed milk
- Poor support by family members at home and the demands of caring for older siblings
- Cultural taboos around public breastfeeding
- A dearth of public breastfeeding rooms

The breastfeeding lobby will never match the deep pockets of Big Formula, but success like the intensive breastfeeding campaign in Madagascar proves the numbers of mothers adopting the practice can be significantly improved: In four years, Madagascar's breastfeeding rate rose from 34 percent to 78 percent.

Meeting the Quintessence Breastfeeding Challenge in Wangjiang Park, Chengdu, Sichuan, China. © Yushi Zhang 袜子

In some countries, the breastfeeding rate is in the single digits. In many parts of the world, this natural process needs reintroduction.

Global lactivism is on the rise. WABA runs an annual World Breastfeeding Week. The Global Synchronized Breastfeed is another global event. As of 2013, the Philippines holds the *Guinness Book of World Records* title for the most breastfeeding mothers nursing simultaneously at multiple sites—31,000.

In 2012, Kelly Schaecher nursed her baby at a trendy café in Bristol, United Kingdom. When a rude server asked her to leave, she posted the incident on Facebook; 250 nursing mothers quickly flash mobbed the café in protest.

Chinese mothers have celebrated National Breastfeeding Day by mobilizing flash mobs, too, raising the profile of nursing moms. Breastfeeding has made a comeback in China after 2008, when poisoned formula sickened hundreds of thousands of babies, some of whom died. UNICEF/China has introduced an app listing public nursing nooks and invites mothers to rate them. Breastfeeding has eco-benefits, too:

WHO recommends that HIV-infected mothers breastfeed, but they should take antiretrovirals (ARVs) and receive guidance from health experts so their babies will be HIV-free; see more in ARVs (**#13**).

- ✔ Carbon neutral: Breastfeeding requires no packaging or shipping.
- ✔ No air-polluting emissions; no fuel is consumed to boil water used to mix the formula or clean the bottles.
- ✔ No deforestation (no wood cut down for fuel to boil water).

How to incentivize breastfeeding? Great Britain is providing low-income mothers who agree to breastfeed with food vouchers, essentially subsidizing the practice. Results of this experiment are not yet in, so stay tuned.

YOU

- Support **BabyMilkAction.org**, a British NGO monitoring false and misleading formula advertising worldwide. They provide supporters with information about their campaigns to identify abuses and organize responses.
- Help UNICEF advocate for additional countries to adopt or strengthen legislation.
- Organize a Breastfeed-In through the Quintessence Breastfeeding Challenge. **www.babyfriendly.ca**
- Submit your best breastfeeding photograph to the World Breastfeeding Week contest.

EDU

- Baby Milk Action is developing an online training course on monitoring the baby food industry.

Assata Doumbia and a friend provide kangaroo care for her twins at a Save The Children care center, Bamako, Mali. © Joshua Roberts

Kangaroo Care

Kangaroo mother care—placing the baby skin-to-skin against the chest—raises survival rates of premature and low birth weight babies.

2

Worldwide, one in five babies born to low-income mothers are premature or underweight. Contributing factors include maternal malnourishment, infections like malaria, and the absence of prenatal care. In low-income countries, an estimated 90-plus percent of extremely early babies (born at less than thirty-seven weeks) die within their first few days.

Incubators are largely absent in low-resource areas. If present, incubators often sit around unused because there is no available electricity. When parts break, there is often no way to replace them.

In 1983, Colombian neonatologists Edgar Rey and Hector Martinez, frustrated by their lack of resources for saving preemies, experimented with using mothers' bodies as incubators, kangaroo-style. It worked. Their technique has been adopted worldwide.

Kangaroo mother care (KMC)—low-tech, high-impact, and easily taught—significantly boosts survival rates for pre-term babies. For stabilized babies, kangaroo mother care is comparable to an incubator and provides added benefits. Save The Children estimates KMC could save more than 450,000 babies every year.

Mothers snuggle their babies skin-to-skin, with the baby held close by a cloth wrapped around them both. Babies wears just diapers, plus hats and socks to conserve body heat, breastfeeding on demand. Fathers, family members, friends, or volunteers provide kangaroo care, too, minus breastfeeding.

Studies confirm its many benefits, almost like extra time in the womb. The babies:

- ✔ Regulate their metabolisms, helped by the rhythm and warmth of their mothers' bodies
- ✔ Sync heartbeats with their mothers'
- ✔ Breastfeed on demand, ideal nutrition with emotional bonding
- ✔ Have stronger resistance to infection
- ✔ Cry less, thereby expending their energy stabilizing and maturing
- ✔ Gain weight faster, with boosted brain development

WHO endorses KMC as a verified, accepted standard of care, now promoted in middle- and high-income countries as well, complementing and humanizing high-tech neonatal intensive care. Kangaroo mothering doesn't require equipment, but mothers must wear their babies 24/7. The first stage takes place in designated hospital units.

Malawi, with very high rates of preterm births, has expanded its KMC program throughout all its twenty-eight districts. Save The Children has been a leader in KMC initiatives, in partnership with USAID. UNICEF and the Bill and Melinda Gates Foundation also work to extend Kangaroo Care training and practice around the globe.

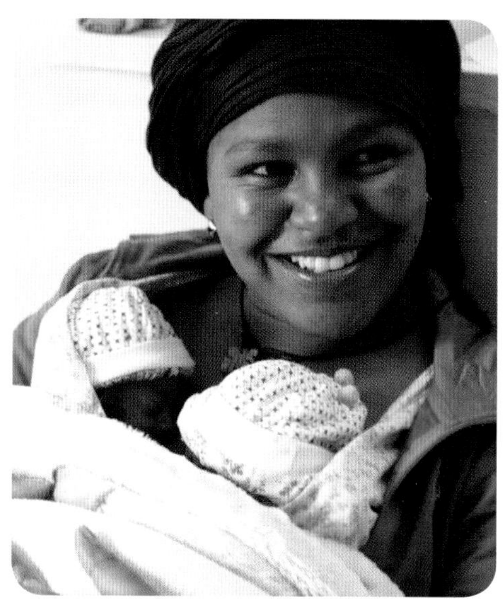

Kangaroo caring mom with her twins, skin-to-skin © UN Foundation/Talia Frenkel

See Also: **Embrace Infant Warmer (#3)**

- **Maternova.net** Research is designing a KMC wrap pattern for seamstresses, creating microenterprise opportunities and local supplies. Maternova serves 170 countries; the pattern and accompanying guidebook will be translated into as many languages as possible.

Embrace Infant Warmer

Embrace Infant Warmers fight hypothermia in preterm and low birth weight babies for 1 percent of the cost of an incubator.

Embraceglobal.org • @EmbraceWarmer • EmbraceInnovations.com

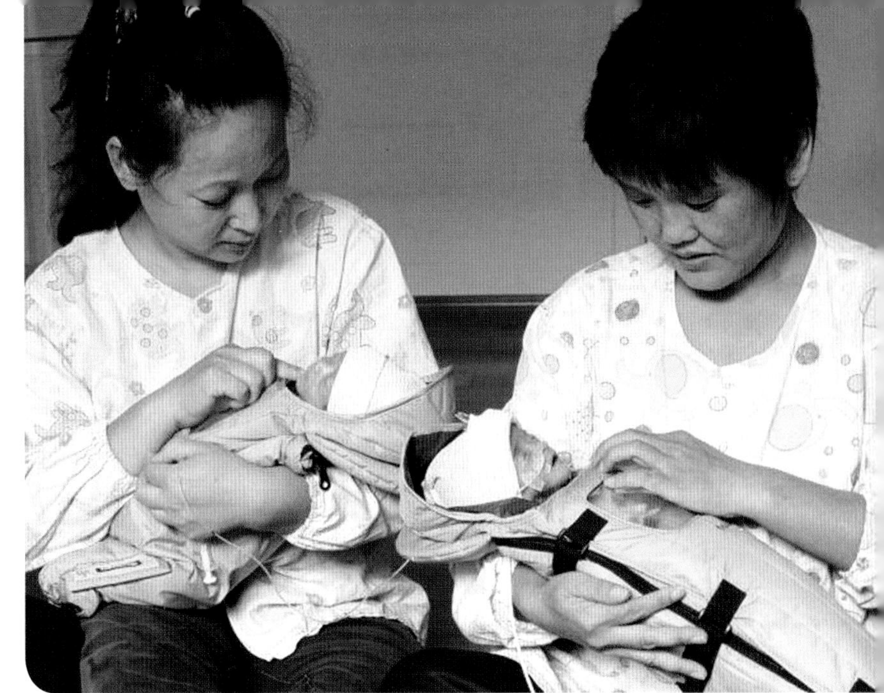

Babies in Embrace Infant Warmers at Chunmiao Little Flower Orphanage, Beijing, China © Chunmiao Little Flower

Embrace Infant Warmers grew out of Stanford University's famous Design for Extreme Affordability course. This rigorous multidisciplinary class brings students together to solve problems faced by the planet's poorest people.

One 2007 student group tackled premature and low-weight birth babies' high mortality rates, especially in the first day of life. *Born Too Soon: The Global Action Report on Preterm Birth* reports prematurity rates, the leading cause of death for children under five worldwide, are rising. Many premature or low-birth weight infants develop hypothermia, the inability to regulate body temperature due to insufficient body fat. Traditional measures like hot water bottles or heat lamps are ineffective.

Their solution is an infant warming bag housing a heated pouch filled with phase-change material. This wax-like substance, once heated, maintains a steady, optimal body temperature of 98.6°F for four to six hours without electricity.

Embrace Infant Warmers provide a locally implemented solution, appropriate for settings where high-tech equipment is often down due to power outages or lack of parts, or is missing altogether. After the course's end, several team members stuck with the project, compelled by its potential to save so many lives. They spent two years in India, learning as much as possible about end users' beliefs and habits, creating dozens of design iterations.

> "It is an injustice that babies continue to die like this all over the world—for a lack of appropriate, affordable technologies."
>
> —*Jane Chen, Embrace co-founder*

Distributing the warmers did not result in their use. The team concluded, therefore, that education is just as important as technology. Now they provide ongoing training and support, demonstrating their warmer and kangaroo mother care (**#2**). Embrace, intended for babies from 3.3 to 5.5 pounds, has come through tests with flying colors; it is credited with savings thousands of babies.

- ✔ It costs roughly 1 percent of the cost of an incubator (under $200 vs. $20,000).
- ✔ As quantities increase, costs will decrease.
- ✔ Its portable design is intuitive and culturally accepted, facilitating more contact with babies than conventional incubators (also called "isolettes").
- ✔ Warmers can be reused up to fifty times, needing just sterilization between uses.
- ✔ Warmers are portable and can be used during transport.

A rural/at-home infant warmer, Care, is undergoing clinical trials in India, a project that is being overseen by Embrace's sister organization, **Embrace Innovations**.

In a promising development, a recent study in Zambia determined that wrapping pre-term babies' torsos with food-grade plastic bags before blanketing them lowered the incidence of hypothermia, without causing hyperthermia (overheating), for 2 cents a bag. The researchers believe the plastic bags prevent evaporative cooling, keeping the babies warm for longer. In settings lacking diapers (**#48**), kangaroo caring mothers can use this plastic bag technique to prevent being soiled by their babies.

Horseback house call by Justina Munoz Gonzalez, vaccinating a baby living near the remote village of San Pablo, Nicaragua. © Adrian Brooks, Courtesy of Photoshare

Vaccine Delivery

Weak vaccination delivery systems fail many of the world's poorest children, who die from preventable diseases. A coordinated global effort to raise vaccination rates is saving many lives.

ShotatLife.org • @ShotAtLife • path.org • @PATHtweets

Children in prosperous countries have been vaccinated against child-killer diseases for generations. Measles, diphtheria, pertussis, tetanus, and polio are vague memories to those living where vaccination is the norm. But millions of the world's have-not children routinely die from these preventable diseases.

Why are so many children not vaccinated? Barriers to providing immunization in low-resource settings include:

- Prohibitive costs for low-income countries' health budgets; end users cannot pay
- Low-functioning health delivery infrastructures
- Lack of electricity for vaccine refrigeration
- Weak demand; vaccines' benefits are invisible and not understood by end users

When Melinda and Bill Gates committed to donating billions to help solve global health problems, they approached this challenge as businesspeople, zoning in on child vaccination as a cost-effective, impactful intervention. The Gates Foundation has leveraged its massive resources not only to underwrite vaccine distribution but to coordinate, cajole, and catalyze all the players to vastly expand poor countries' vaccination programs.

The resulting **Global Alliance for Vaccines and Immunisation (GAVI)** launched in 2000. GAVI brings together donor governments, WHO, UNICEF, the World Bank, the pharmaceutical industry, research and technical agencies, and The Gates Foundation, among other donors.

This partnership has reversed the stagnating, or even declining, rates of vaccination in low-income countries, facilitating the immunization of nearly 400 million children. Behind those numbers are real children whose lives were saved, or who were spared lifelong disabilities—a gift also to the families who would care for them.

Higher child survival rates shape parents' expectations. Demographics show that as people climb out of extreme poverty, they have fewer children. Confident in their children's survival, they invest more resources in each offspring, creating an upward spiral of improved health and prosperity.

Vaccinations administered in the developing world include the old-standby vaccines:

- ✔ Polio*
- ✔ Measles*
- ✔ DPT*—Diphtheria, Pertussis (Whooping Cough), and Tetanus

These have been augmented by more recently approved vaccines:

- ✔ Rotovirus—Protects against a common form of diarrhea.
- ✔ HPV—Protects against *human papillomavirus*, the cause of 99 percent of cervical cancers.
- ✔ Meningitis A—Commonly given in the "meningitis belt" of Africa.
- ✔ PCV—Prevents a common form of pneumonia, one of the biggest killers of low-income children.
- ✔ Hepatitis B*— Protects against the most common form of hepatitis and liver cancer, which can frequently ensue.
- ✔ Hib*—Protects against *H. influenzae* bacteria, potentially deadly in infants and young children.

*The Pentavalent vaccine combines these five immunizations together, administered in three doses. This lowers prices, lessens stress for children, and requires fewer trips for mothers.

For children, life is safer. Communities benefit, too; when more children are vaccinated, the exposure risk for the nonvaccinated decreases. Herd immunity is a well-documented phenomenon whereby the more people who are vaccinated, the fewer people there are to spread infection to unvaccinated people. In this way, large-scale vaccinating programs help prevent infection even among those who are not vaccinated, though this is only effective when a given threshold of vaccination coverage is reached in the community.

Positive vaccination developments include:

- ✔ Lowered costs—Subsidies, along with negotiated agreements with pharmaceutical companies, are radically lowering vaccine prices for poor countries.
- ✔ Technology—Inventions and improvements in delivery tech, including unrefrigerated storage, are expanding distribution.

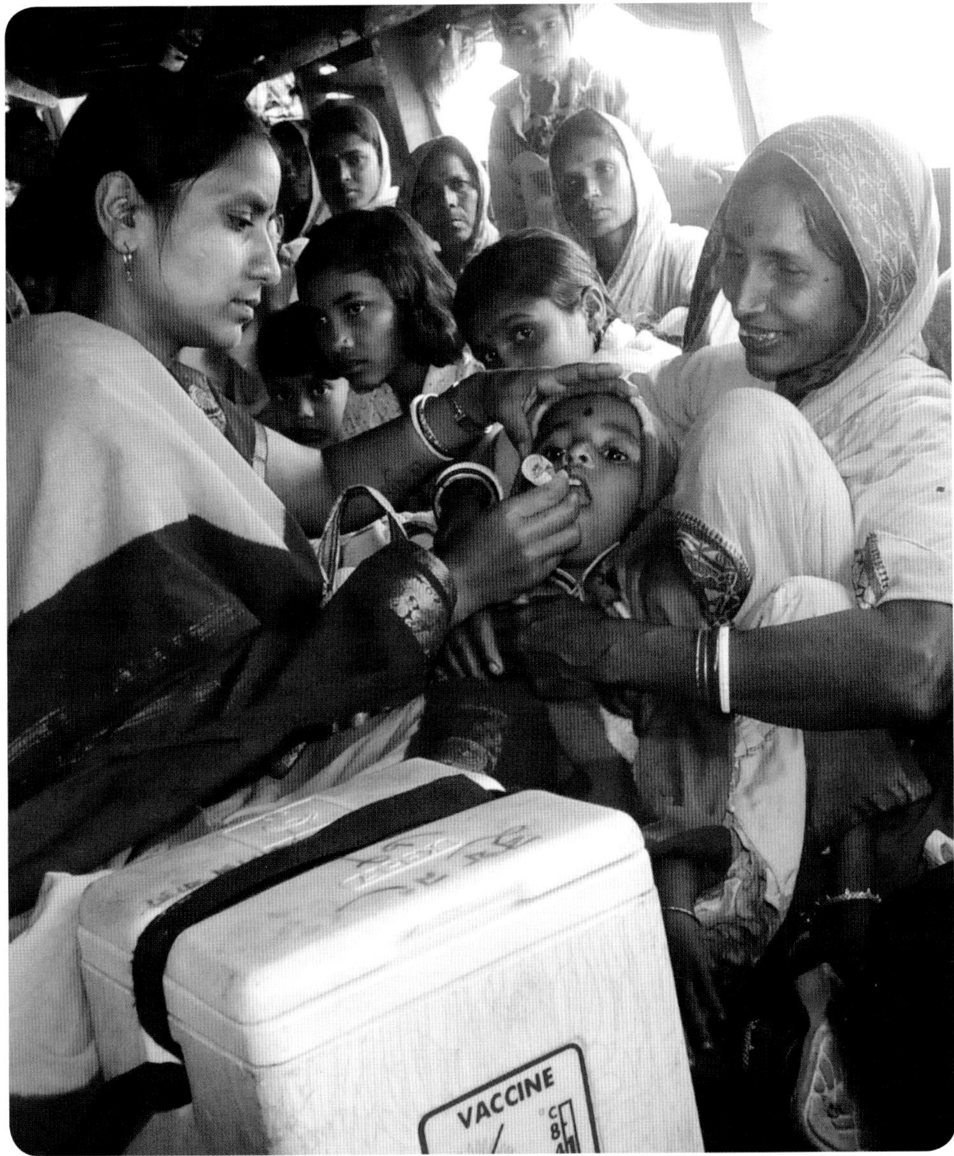

Health worker outreach: Providers administer oral vaccinations on the ferry between the mainland and the islands of Sunderbans, West Bengal, India. © Tushar Sharma, Courtesy of Photoshare

- ✔ Awareness Campaigns—Public education and the support of local leaders helps increase acceptance of the value of vaccinations, fighting mistrust and disinformation.
- ✔ Outreach—Creative initiatives help immunize more children. Texting pregnant women and new mothers, for example, reminds them when upcoming shots are due.

Esther Duflo is a pioneer in the field of poverty economics. Her study near Udaipur, an area of India with an immunization rate of just 2 percent, compared approaches to increase vaccinations. She found that public health workers showed up at clinics only about half the time, so that when motivated mothers did make it to the clinic, it was often in vain. The study also found that:

- ✔ Setting up immunization camps with guaranteed personnel increased rates to 18 percent.
- ✔ At other sites, parents were given a kilo of lentils at each vaccination visit and a set of plates when the series was complete; vaccination rates rose to 38 percent—and the increased efficiency offset the cost of the lentils.

Creative approaches based on studies like this, capitalizing upon lessons learned in behavioral economics, can expand the success of vaccination initiatives' success.

Where women have low status and limited decision-making power, rates of immunization of both sons and daughters is lower. Immunizing even one child improves the family's health, since sick children drain family resources and infect siblings.

The new HPV shot prevents cervical cancer (**#17**), a potentially lethal disease caused by the sexually transmitted human papillomavirus (HPV). HIV-positive women are especially vulnerable, but all women are potentially at risk. About 300,000 women die from cervical cancer annually, almost all in low-income countries. The HPV vaccine will save many women's lives and prevent hundreds of thousands of children from becoming motherless.

HPV vaccines are given to girls aged nine to thirteen; girls that age typically have little interaction with low-income countries' health services, a challenge to overcome. The HPV vaccine is available in affluent countries, too.

See also: **Deworming (#10)**

- If your daughter is nine or older, make sure she receives her HPV vaccine. The CDC also recommends the HPV vaccine for boys aged eleven or twelve.

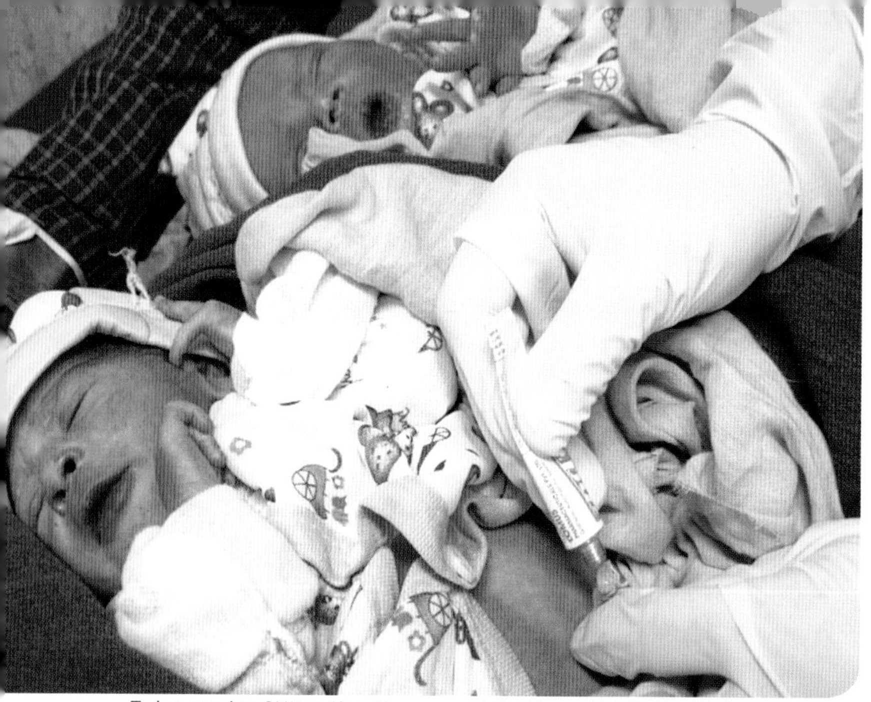
Twins receive CHX antiseptic cream at the Nepal Family Health Project.
© Meena Suwal/John Snow, NFHP

Umbilical Cord Treatment

A single 25 **cent** application of CHX gel on the umbilical cord stump can save a baby's life.

HealthyNewbornNetwork.org • @HealthyNewborns

Many babies born in places lacking resources to maintain germ-free environments die from bacterial infections contracted in their first days of life. **Chlorhexidine**, a common mouthwash and pre-surgical prep antiseptic, also protects against infections entering newborns through their umbilical cord stumps. One gel smear on the umbilical cord is all it takes.

CHX is manufactured in-country for 25 cents a tube. The drug costs a cent; the balance pays for packaging.

Half a million newborns die annually from acquired infections, 99 percent in low-income countries. Randomized clinical trials in Bangladesh, Nepal, and Pakistan showed CHX lowers infant mortality from infections by 25 percent to 66 percent. These are some of the largest impacts ever reported for neonatal interventions.

> "Chlorhexidine for umbilical cord care: A 'best buy' for newborn health."
> —*HealthyNewbornNetwork.org*

Besides its low price tag, CHX has many added selling points:

- ✔ A long-term safety record around newborns
- ✔ Common availability worldwide
- ✔ Long shelf life, no refrigeration required
- ✔ Simple application, requiring no training or equipment
- ✔ Long-lasting adherence to babies' skin, prolonging effectiveness
- ✔ Local acceptance of both liquid and gel form

Dry-cord treatment, simply leaving the stump to heal on its own, has been the prevailing medical recommendation, a widely ignored directive. People sense that a baby's drying umbilical cord is vulnerable; human instinct is to protect babies by applying substances to their cord stumps. Some applications are meant to promote drying, like charcoal. Another approach is lubricating with mustard oil or rubbing on dung or turmeric.

It is likely that some folk treatments actually introduce infections, making the promotion of CHX gel a culturally sensitive problem-solver. The Bangladesh/Nepal/Pakistan study affirmed that substituting CHX for traditional treatments was widely accepted.

Nepal, a low-income country with many home births and a correspondingly high infant mortality rate, was the first to adopt countrywide CHX distribution. Sokoto State in Nigeria has also moved to procure CHX for its health system. In 2013, the World Health Organization added CHX to its Model List of Essential Medicines for Children. This will likely prompt many more countries to adopt CHX postnatal treatment.

Delivering CHX tubes to where they are most needed, at home births, is challenging. Women give birth at home because they live in remote, low-resource areas—precisely where supply chains, distribution systems, and public health networks are weakest.

See also: **Clean Birth Kits (#20)**

YOU

- CHX is an excellent **cause marketing** candidate. Partner with an NGO to distribute CHX, donating a dose for each item sold. **Babies4babies.com** donates one CHX dose per purchase; customers receive a code that, when entered, funds a second dose.

- Students of design, public health, and marketing: Create country-appropriate packaging designs and distribution campaigns for CHX uptake. Many low-income countries are interested in manufacturing and distributing CHX.

Tooth Brushing

Simple tooth brushing helps reverse the pandemic of children's tooth decay caused by junk food and sugared drinks.

GlobalGrins.org • @GlobalGrins • smilesquared.com • @Smile_Squared

A Guatemala clinic with University of Pennsylvania nursing students. © Smile Squared/Stephanie Bosch

6 When pediatrician and med school professor Dr. Karen Sokal-Gutierrez revisited Central America, near where she served in the Peace Corps, she was shocked to see young children's smiles revealing blackened stumps. She was sure that had not been the case a generation ago.

Back at home, she checked her 1970s photos; those pictures featured children with pearly teeth. The culprit? Cheap, processed starchy and/or sugary snacks and beverages. The arrival of junk food ahead of tooth brushing brings an onslaught of cavities.

Advanced tooth decay makes eating painful and disrupts school concentration, play, and sleep. Children with diseased and missing teeth suffer from embarrassment and humiliation.

Many factors are behind the decay epidemic:

- ✗ When families move out of extreme poverty, parents can give their children spending money, which they spend on candy.
- ✗ Junk snacks are often cheaper than healthier options, a worldwide problem. Where drinking water is unsafe, sugary drinks are often cheaper than bottled water.
- ✗ Snack food marketers target children directly, using cartoon characters and bright, shiny packaging.
- ✗ More kids are enrolled in school worldwide; junky snacks are peddled adjacent to schools.
- ✗ Many parents are unaware that diet is causing the tooth decay, and that tooth brushing and decreasing sugar intake prevents it.
- ✗ Dental care and education is not integrated into low-income countries' public health systems.

In 2004, Sokal-Gutierrez mobilized a response, initiating UC Berkeley's **Children's Oral Health and Nutrition Project** in collaboration with local NGOs in El Salvador. Every summer, a team spends two weeks providing on-site dental care and educating local public health workers about the importance of dental hygiene, especially for children under age six, with impressive results. Locals have integrated dental care and education, and kids' teeth are now well cared for. Alums have exported the program to Ecuador, Peru, Nepal, India, and Vietnam.

Helpful interventions include:

- Distributing toothbrushes; many NGOs already do this. Some schools include tooth brushing during the school day.
- Regulating marketing to children to help reduce demand.
- Taxing junk food and sugared drinks, following Mexico's lead, providing revenue for public health programs, including dental hygiene initiatives.
- Marketing inexpensive fluoridated toothpaste. If candy is affordable, fluoridated toothpaste designed and marketed to children should also be affordable.

YOU

- For each eco-friendly **Smile Squared** toothbrush purchased, one is donated to a child in a low-resource region.
- Travelers, **Global Grins** will provide you with a box of 100 toothbrushes to personally deliver to an orphanage, clinic, shelter, school, or other organization.
- When traveling, do not give children candy as a gift. Sokal-Gutierrez recommends pencils or small toys.
- Dental professionals can research volunteer opportunities through the American Dental Association's portal, **InternationalVolunteer.ada.org**.

EDU

- Students should check out **GlobalBrigades.org**.

Rajastan Self Help Group produces Raj-Nutrimix (#18), a micronutrient-fortified supplement. © Global Alliance for Improved Nutrition (GAIN)

Micronutrient-Fortified Food and Supplements

Fortifying staple foods with essential micronutrients, an inexpensive process, prevents blindness, mental retardation, and maternal deaths caused by micronutrient deficiencies.

micronutrient.org • @micronutrient • GainHealth.org • @GAINalliance • Vitaminangels.org • @VitaminAngels

7 The majority of countries mandate **food fortification**, adding trace amounts of essential micronutrients during food processing. This prevents populations from the maladies caused by micronutrient deficiency. Meager diets may provide enough calories for survival, but their nutritional deficiencies sap energy and impair health.

An estimated 30 percent of the world's population lives in countries lacking food fortification. Women and children in extreme poverty disproportionately suffer from nutrient deficits, so they stand to benefit significantly from food fortification, a well-established intervention. The Copenhagen Consensus, a gathering of economists and Nobel laureates recommending anti-poverty priorities, declared vitamins for undernourished children "the world's best investment."

> "When vitamins and minerals are added to wheat flour, maize products, and rice, commonly eaten foods become more nutritious. Consequently, consumers improve their health without changing their habits. The extra nutrition helps people become smarter, stronger, and healthier."
>
> —*Flour Fortification Initiative*

Vitamin A deficiency:

✗ Depresses children's immunity
✗ Can cause vision impairment, including blindness, in children six months to five years old

When malnourished kids receive vitamin A supplements, their resistance to infection is strengthened, lowering their death rates from malaria and diarrhea. The solution is twice yearly slow-release vitamin A capsules, costing 2 cents a dose.

Iodine deficiency results in:

✗ Fetal mental retardation
✗ Brain damage in young children
✗ Goiter, an extreme swelling of the thyroid gland
✗ Stunting of growth

The highest priority is to provide iodine to mothers of childbearing age, especially pregnant women, and young children. The solution is iodized salt used normally year round, costing 5 cents per person per year, with USAID reporting a return on investment of $28 per dollar spent.

Iron deficiency causes anemia, which:

✗ Impairs infants' mental development, permanently lowering IQ
✗ Saps energy from nearly 500,000 women worldwide (**#19**)
✗ Endangers pregnant women and their fetuses, causing an estimated 50,000 maternal deaths annually, plus low-weight babies at high risk (**#2**)

Solutions include fortifying flour, other grains, oils, and condiments like fish sauce, costing 10 cents per person per year. Return on this investment: $84 per dollar spent.

> **Double Fortified Salt,** iron + iodine, is a newly introduced product.

Folic acid deficiency causes fetal spina bifida and other neural tube defects, as well as anemia. Supplements are important for adolescent girls and women, especially if they are pregnant. Fortifying flours and cereals with folic acid, along with iron, costs 10 cents per person per year.

Multiple micronutrient supplements combine all of the above:

✓ **Ultrarice®**, developed by Path.org, is a micronutrient-enriched faux-rice grain mixed 1 to 100 into regular rice.
✓ **Sprinkles™**, produced by SGHI.org, is a micronutrient packet added to children's food.

See also: **Prenatal and Postnatal Care (#18), Nutritionally Enriched Crops (#59)**

YOU EDU • Start an advocacy initiative by accessing the **Global Hidden Hunger Indices and Maps** (online).

School Lunch

Free school lunches improve students' attendance and performance and are a proven incentive for educating daughters.

FoodForEducation.org • @AkshayaPatraUSA • MarysMealsUSA.org

Bulgur and lentils for lunch, Sierra Leone. © Lane Hartill/Catholic Relief Services

Research shows free in-school lunches, snacks, and take-home rations:

- ✔ Improve children's abilities to concentrate, learn, and perform tasks
- ✔ Increase girls' school enrollment
- ✔ Improve attendance
- ✔ Have the highest impact on chronically undernourished children
- ✔ Support local farmers and create jobs in food service
- ✔ Provide a channel for additional health interventions

Programs often target girls, enabling them to access education in societies that traditionally under-invest in daughters and are disinclined to send them to school. Food parcels sent home with students promote attendance, reaching the most food-insecure families.

The **UN's World Food Programme**, funded by governments, corporations, and individuals, is the world's largest lunch provider. They serve more than 20 million students daily, costing around $50 a year per student.

Akshaya Patra feeds more than 1.3 million Indian primary school students daily through its public/private partnership. Using local farmers and vendors lowers transportation and food costs while supporting local economies. **Akshaya Patra USA** supports the India program through chapters in more than twelve American cities.

The Advancement of Rural Kids, **ARK**, integrates eco-agriculture into the Philippine schools they support. Parents, staff, and students learn sustainable agriculture techniques, growing more than nineteen kinds of vegetables in their cooperative garden. Yields contribute to lunch preparations.

The Kasiisi Porridge Project provides lunches at the Kasiisi School in Western Uganda near Kibale National Park. Anita, a twelve-year-old student there, comes from a family that, like most families in this area, eats one daily meal. After fetching water and sweeping the family's compound, she and her friends walk to school hungry. It seems that porridge packs a punch; she landed a regional secondary school scholarship, hoping to one day become a doctor.

In 2012, the photo on the blog of a nine-year-old Scottish school girl, Martha Payne, went viral. The picture on her blog, which was titled *Never Seconds*, featured her skimpy school lunch. She leveraged the ensuing media attention, and partnered with **Mary's Meals**, a Scottish NGO that provides school lunches in sixteen countries. Martha works with their Malawi program.

Enjoying soup made from vegetables grown in the school garden, Bitoon Ilaya, Philippines. © Mabini Abellano/Advancement for Rural Kids

Challenges:

- Funding
- Avoiding corruption in food sourcing and preparation
- Providing nutritious, child-friendly food

- **Akshaya Patra USA** is eager to utilize volunteers' professional skills, such as photography, blogging, grant writing, and graphic design.
- **Akshaya Patra** offer tours of their state-of-the-art kitchens; visit and lend a hand when you are next in India.
- **KasiisiPorridge.org** welcomes on-site volunteers.

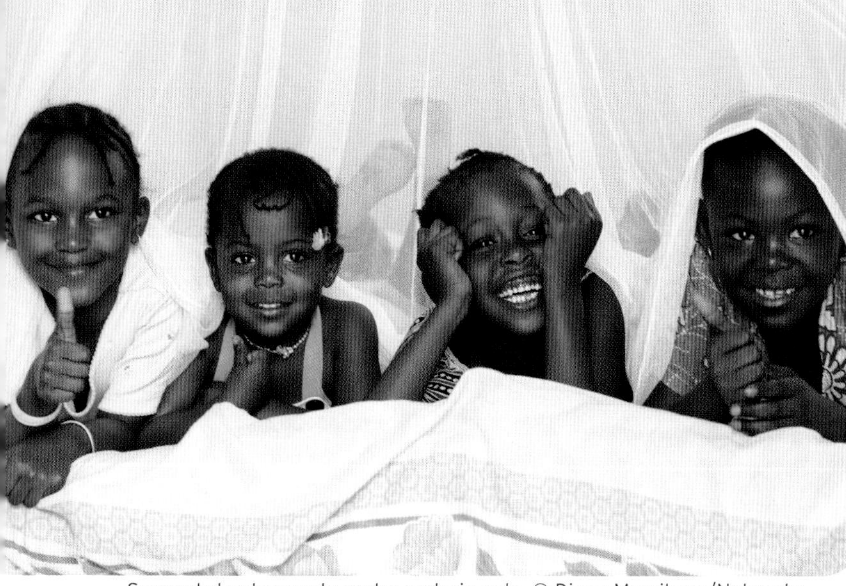
Senegal slumber party under malaria nets. © Diana Mrazikova/Networks, Courtesy of Photoshare

Malaria Prevention and Treatment

Insecticide-treated nets prevent malaria. Inexpensive, quick diagnosis and treatment saves lives.

H tel

nothingbutnets.org • @nothingbutnets • MalariaNoMore.org • @MalariaNoMore

Malaria, an infection causing raging fevers, can kill. Treatment—quinine from tree bark—was discovered in the 1600s. The cause, bites from disease-carrying mosquitoes, was identified in the late 19th century. Despite being preventable and treatable, it is the third-highest cause of children's deaths globally. About 90 percent of cases are in Africa, but it is also present in Asia and Latin America.

Combating malaria poses big challenges. Where commonplace, it is not perceived as an emergency; most people suffer with it and recover. Educating and promoting prevention is crucial to malaria eradication.

- ✗ Malaria kills. There are an estimated 300 million cases annually. With a fatality rate of less than one-tenth of 1 percent, that's still a million deaths, mostly children.
- ✗ Malaria weakens survivors. Repeated rounds of infection can cause severe anemia, increasing patients' vulnerability to other illnesses.
- ✗ Malaria during pregnancy can cause pregnancy loss or premature or low-weight babies with low survival odds. Many pregnant malaria patients are already anemic and/or HIV-positive, conditions worsened by malaria, costing tens of thousands of mothers' and babies' lives annually.
- ✗ Malaria saps economic productivity, not only of patients, but also of their caretakers—most of whom are women.

Long Lasting Insecticidal Nets, effective prevention tools, cost under $5. Suspended over beds, they keep out mosquitoes. With correct use, infections are cut in half, with 20 percent fewer children's deaths.

Bed net use, estimated at below 50 percent, is difficult to measure. Many factors decrease compliance:

- It's hard to get excited about nonevents; nets' protective value is abstract.
- Nets partially block air circulation, a discomfort in hot regions.
- Nets can develop holes or rips, and after four to five years the insecticide wears off.
- Sometimes mosquitoes are trapped inside the nets.
- Recipients may lack commitment to using free nets.

Some programs require end users to pay a token fee, increasing buy-in and compliance. People still catch malaria, despite precautions. Properly equipped frontline clinics are central in fighting malaria, now that affordable diagnosis and treatment exists.

- ✔ Rapid Diagnostic Test Kits cost $1, with results in fifteen minutes.
- ✔ State-of-the-art malaria treatment, **artemisinin-based combination therapies (ACTs)**, costs $1, curing children in one to three days.

Coordinated efforts by governments, NGOs, and international health organizations have lowered death rates 25 percent over the last decade. Tanzania has slashed malaria deaths for children under age five by 45 percent. When malaria cases decline, so do deaths from other causes, a win-win.

Two African inventors, Moctar Dembele of Burkina Faso and Gerard Niyondiko of Burundi, won the 2013 Global Social Venture Competition for their invention, **Fasoap**. A mosquito repellent made of shea butter, lemongrass, and additional secret local ingredients, it is undergoing further testing.

Research on malaria vaccination is ongoing.

- The **Against Malaria Foundation** recruits volunteer video editors to condense footage from each bed net distribution.

Deworming

Lucienne, a health and social worker at the Saksida Child Development Center, administers deworming pills, Burkina Faso. © Compassion International

A vast majority of children in low-resource areas suffer from intestinal worms. Deworming capsules improve health and school attendance at the cost of 50 cents per child.

DewormTheWorld.org • @EvidenceAction • PovertyActionLab.org • @JPal_global

Intestinal worms are parasites that take up residence in involuntary hosts. In affluent settings, people are familiar with their pets being afflicted. Worms are a troublesome human ailment in low-resource areas, causing:

✗ Malnourishment
✗ Chronic anemia
✗ Diminished energy
✗ Increased vulnerability to other infections

Parasites thrive in warm climates in areas lacking modern sanitation. Soil-based parasites are transmitted via bare feet. Some spread through fecal matter. Water-borne parasites attack bathers and swimmers.

People are often unaware they are infected. Worm-caused discomforts are so prevalent, many people consider them normal. Treatment has long been available, but not widely administered.

MIT's **Poverty Action Lab**, known as **J-PAL**, scrutinizes interventions, identifying the most effective poverty alleviators. The 2004 publication of a randomized study by J-PAL affiliates Michael Kremer and Edward Miguel showed deworming improved health and school attendance. J-PAL recommends deworming as a "Best Buy."

The study estimated that deworming in communities with heavy infection loads results in 25 percent fewer school days missed. Because education has a high return on investment, deworming offers large payoffs. A decade later, recipients of deworming treatment work more hours and eat more meals.

Deworming initiatives are generally school-based, more cost-effective for reaching children than working through health clinics. Teachers are trained to administer the medications and maintain records, further lowering overhead.

Because it is cheaper to treat worms than diagnose them, schoolchildren living in areas with high infestation levels are given two medicines:

✓ **Albendazole** or **Mebendazole** treats all common intestinal parasites (helminths)—hookworms, roundworms, and whipworms. One size fits all, given twice yearly.
✓ **Praziquantel** treats schistosomiasis (bilharzia), water-borne worms—doses depends on weight, given once yearly.

Deworming Days can be integrated with vitamin A capsule distribution (**#7**) and vaccinations (**#4**), for greater synergy and cost-effectiveness.

> "The most disadvantaged children—such as girls and the poor—often suffer most from ill health and malnutrition, and gain the most benefit from deworming"
> —*SchoolsandHealth.org*

It is important to reach out to those missed by school-based dewormings, such as children not enrolled in school. Treating pregnant girls is vital. Studies show that in vulnerable communities, deworming reverses maternal anemia and improves birth weight and child survival.

Adults suffer from worms too. When schoolchildren are dewormed, untreated community members also benefit since the number of transmitters drops.

To date, there has been no evidence that parasites afflicting humans are becoming drug-resistant, though this has been observed in animal deworming.

Challenges:

- Programs need funding; when schools charge for treatment, participation plummets.
- Distribution infrastructures need to be developed and maintained.
- Trust needs to be developed. Generally parents get on board when they observe their kids shaking off symptoms.
- The underlying cause of worms is poor sanitation, covered in Sector 4.

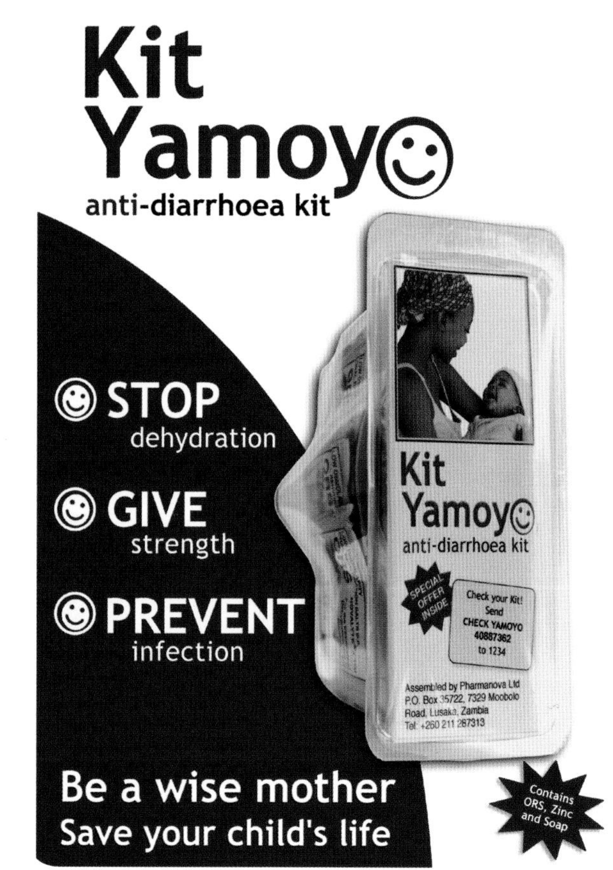

Kit Yamoya, ColaLife's Oral Rehydration Therapy prize-winning package. © Guy Godfree

Oral Rehydration Salts + ColaLife

Oral Rehydration Salts have saved millions of lives by preventing the dehydration caused by diarrhea in malnourished people.

colalife.org • @ColaLife

The recipe for **Oral Rehydration Salts (ORS)**, a premier 20th century health innovation, is simple:

- ✔ 6 T. sugar
- ✔ ½ t. salt
- ✔ 1 liter of clean water

Diarrhea annually kills more than 1 million malnourished children through its lethal weapon: dehydration. ORS helps the gut absorb water, saving lives. ORS was introduced in 1971 to treat cholera during the Bangladeshi War of Independence. It dropped the fatality rate from 30 percent to 1 percent.

Health clinics in low-resource areas provide ORS in packets. Affluent parents reach for a Pedialyte bottle, a premixed glucose/electrolyte solution—sugar, salt, and sterilized water. In high-resource settings, diarrhea is easily treated. Low-resource areas face huge treatment distribution hurdles. Simon Berry, an affable British humanitarian health project manager working in Zambia, long marveled at Coca-Cola's penetration in even the most remote of locations. Was there a way to link distribution of vital medicines like ORS to Coke deliveries?

In 2011, Simon and his wife, Jane, posted their vision for an improved ORS distribution initiative on Facebook. People loved their idea for attractive kits catching a ride with Coke.

A kit includes:

- A reusable, resealable container
- Eight single-dose ORS sachets calibrated for toddlers to sip over a 24-hour period
- Zinc tablets, for added diarrhea treatment
- A bar of soap, promoting prevention
- A cover to keep out dirt and flies

The resulting project, **ColaLife**, designed **Kit Yamoyo** ("kit of life") to fit into the negative spaces of crated Coca-Cola bottles. Designed by PI Global, this award-winning package has been very successful in trials. Kiosk vendors purchase kits along with Coke and other sundries, netting a small profit on each sale. The first purchaser was a distributor—his grandson had a bad case of diarrhea.

Demand is strong. Moms with sick children want to do the right thing; Kit Yamoyo helps them treat diarrhea more confidently. Children in low-resource areas average three diarrhea episodes a year, so the package's coupon offering a discount on a future purchase is very popular. The cost is modest, equivalent of the price of five bananas.

Coca-Cola supports the project without direct involvement. Since its international structure is decentralized, administrative decisions are left to national-level management.

ColaLife consults extensively with end users, demonstrating that well-designed, branded, affordable products can save kids' lives. You can follow their progress on their highly informative blog.

- Sponsor a screening of Claire Ward's prize-winning documentary about ColaLife, *The Cola Road*. –Claire.Ward@nyu.edu
- Take ColaLife to another region. ColaLife is an **open-source** project. Each country needs its own version, connecting NGOs, health ministries, donors, and the national Coca-Cola bottling network.
- Market ColaLife in the USA, or another upscale market, to subsidize ColaLife's humanitarian sales. Its eco-friendly packaging is greener than present products, and consumers like cause-related marketing.

Plumpy'Nut™: Ready-to-Use Therapeutic Food

Peanut-based Ready-to-Use Therapeutic Food (RUTF) has revolutionized the care of acutely malnourished children, saving vastly more lives than was previously possible.

PlumpyField.com • @PlumpyField • EdesiaGlobal.org • @EdesiaGlobal • ValidNutrition.org

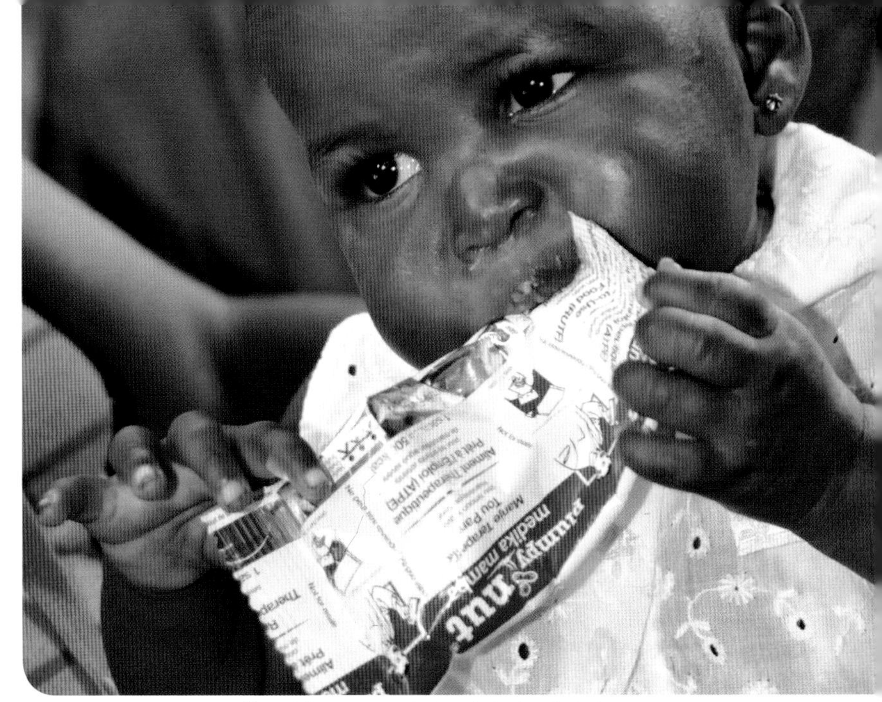

Enjoying Plumpy'Nut™, Haiti. © Navyn Salem/Edesia Global

The world's poorest children are chronically undernourished. Living in areas without safe water, kids are often fighting multiple infections. Children have very little reserve; they can quickly cross the line between "routine" undernourishment to severe acute malnutrition.

At that point, emergency medical intervention is required to save them. Before the invention of **Plumpy'Nut™**, high-energy peanut paste fortified with essential vitamins and minerals (**#7**), acutely malnourished children needed hospitalization. Only a tiny minority made it to such a setting; most died.

UNICEF is now the largest purchaser of Plumpy'Nut™. Together with international NGOs, more than 2 million children were fed through its humanitarian feeding programs in 2012, a record.

Plumpy'Nut™ and other RUTFs (ready-to-use therapeutic foods) have saved the lives of many more acutely malnourished children than was previously possible. RUTFs are prescribed by medical personnel, but hospitalization is replaced with outpatient care. Plumpy, as it is nicknamed, has many positive attributes:

- ✔ Kids like it.
- ✔ Acutely malnourished children can digest and absorb its nutrients.
- ✔ Children eat it independently, eliminating the risks and expense of hospitalization.
- ✔ Mothers (or other caretakers) do not need to stay near hospitals or feeding centers.
- ✔ It allows more efficient deployment of health workers.
- ✔ Individually packaged, it is safe and easy to transport and store.
- ✔ It has a long shelf life, requiring no refrigeration.
- ✔ It is not mixed with water, eliminating contamination risks.

French pediatric nutritionist André Briend conceived the idea in 1999 while observing French children relishing Nutella breakfasts and snacks. He collaborated with Michael Golden, an Irish nutritionist, to create 500-calorie squeeze-packets formulated to be maximally digestible and nutritious, at about 50 cents each. Treatment protocol is twice daily for 7 weeks.

Plumpy'Nut's parent company, Nutriset, is based in Normandy, France. Balancing the desire to reach as many children as possible with strict brand quality control, Nutriset has developed a franchise program. **PlumpyField** network factories are located in:

Burkina Faso	Niger
Dominican Republic	Sudan
Ethiopia	Tanzania
Haiti	United States
India	Uganda
Madagascar	

In-country production provides jobs, supports local farmers (many of whom are female), builds capacity, and lowers costs. In some other countries, factories produce generic versions.

> Childhood peanut allergies are unusual in the Global South. "Food allergy seems far less common in poor countries than in rich countries. This well-known observation has been explained by different factors, but apparently, crowding and repeated exposure to infections seems to play a role." *—André Briend*

YOU
• Capitalize on Plumpy's popularity—in some venues there is a thriving black market for RUTFs—by developing and marketing a nontherapeutic version, generating funds to subsidize the real thing. (Plumpy'Pal?)

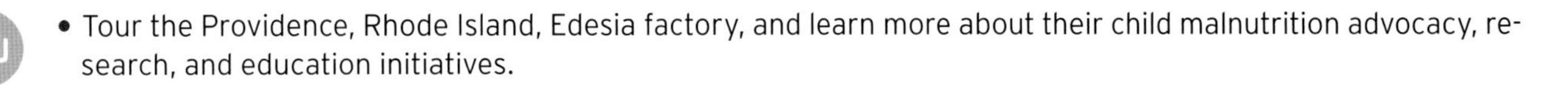

EDU
• Tour the Providence, Rhode Island, Edesia factory, and learn more about their child malnutrition advocacy, research, and education initiatives.

Sister Sade counsels a patient at the Makele AIDS clinic, Ethiopia.

Antiretroviral Drugs: Extending Life with HIV/AIDS

Though antiretrovirals extend the lives of HIV-positive patients, two-thirds of people who need them do not receive them.

BornFreeAfrica.org • @BornFreeAfrica • AMFAR.org • @amfAR

Three decades later after its emergence, there is still no cure for HIV/AIDS and vaccine researchers have yet to score a victory. But there has been progress. Due to ARV drug treatment, many people with HIV/AIDS live much longer lives.

Antiretroviral (ARV) drug therapy, developed in the 1990s**,** extends HIV-positive patients' lives indefinitely**.** It was too costly, though, for the millions in middle- and low-income countries who needed them. Since 2003, negotiated lower prices (partly in response to pressure from AIDS activists), donors, and international aid have made it possible to distribute ARVs much more widely.

ARVs prevent the virus from developing into full-blown AIDS. Without ARVs, that generally occurs within five to ten years. Middle-income countries now manufacture ARVs, increasing supply. Costs of ARVs continue to decline.

> "Investment in AIDS will be repaid a thousand-fold in lives saved and communities held together."
>
> —*Dr. Peter Piot, former Executive Director, UNAIDS*

- ARVs reduce the body's viral load, preventing progression to AIDS.
- Most regimens combine three drugs in a single pill, taken daily.
- Medications are dispensed at health clinics, generally for free, a month's supply per visit.
- Uninfected partners can take ARVs to prevent acquiring HIV/AIDS.
- Medications must be taken with safe water, on a full stomach—challenging for food-insecure people living in settlements lacking sanitation and treated water.

The female face of HIV/AIDS:

- Women are more biologically susceptible to acquiring AIDS from heterosexual sex than men, and comprise 60 percent of HIV cases.
- Women and girls often acquire HIV/AIDS via nonconsensual or transactional sex where they have little negotiating power.
- Infected men often refuse condoms.
- Females typically partner with men who are older and more likely to be infected.
- Pregnant mothers pass the virus to their offspring through pregnancy, childbirth and breastfeeding. (ARVs can now prevent this*.)
- Women generally serve as primary care providers for sick family members.
- Many grandmothers and other female relatives have become full-time parents to children orphaned by AIDS.

Challenges:

- Many barriers remain to extending ARVs to all who need them:
- Costs of medication are high, generally borne by governments and donors.
- High-functioning health infrastructures are needed to administer patients' drug regimes.
- Providing treatment to remote areas far from public health outlets is difficult.
- Adherence is burdensome for people with unstable life circumstances.

Ignorance, fear and social stigma discourage testing and treatment. Robust public health education efforts promote prevention. The central messages are:

- Sexual intercourse is the most common route of infection.
- Condoms prevent infection when used correctly and consistently.

Voluntary circumcision is a proven prevention strategy, lowering the risk of male infection by about 60 percent. **PrePex** is an inexpensive nonsurgical tool increasing access to the procedure. Female condoms also prevent infection.

***BornFreeAfrica** is an initiative whose mission is ending mother-to-child transmission of HIV.

Solar Ear: An Affordable Hearing Aid with Rechargeable Batteries

Using sign language, Sarah and Akangang of the Solar Ear project in Botswana discuss how the solar charger works. They helped develop the chargers and now teach others how to manufacture them. © SolarEar

90 percent of the world's 360 million hearing-impaired people do not have access to hearing aids like the affordable Solar Ear, powered by solar-charged batteries.

tel

solarear.com.br • @EarsForYears.org • WWHearing.org • @WWHearing

About one in twenty people in the world is hearing-impaired. Hearing loss is more prevalent in low-income countries due to the absence of preventative health interventions. Deaf education programs are exceedingly rare in these regions, leaving many hearing-impaired children isolated and illiterate. Hearing-impaired adults face many burdens, including high unemployment.

About 95 percent of hearing-impaired people around the globe would benefit from hearing aids, but costs are prohibitive in low-resource areas. Batteries generally last only a week, so even if people are given hearing aids, they cannot afford the cost of batteries. Many residents of affluent countries are also priced out, since hearing aids typically cost a few thousand dollars, usually out of pocket. Senior citizens who live on fixed incomes, especially, are often unable to purchase hearing aids.

This massive market failure creates an opportunity ripe for innovation. An affordable hearing aid powered by low-cost batteries would improve educational and employment opportunities for hundreds of millions of children and adults.

Solar Ear, a social enterprise founded in Botswana with the assistance of a Canadian, Howard Weinstein, has designed just such a hearing aid at about one-tenth the cost of conventional devices. To power it, they designed solar-charged hearing aid batteries. Instead of lasting just a week, their solar hearing aid batteries can be recharged daily at no cost, and last two to three years.

Solar Ear has three primary missions:

- ✔ Designing and manufacturing affordable products for the deaf
- ✔ Employing hearing-impaired people
- ✔ Creating awareness about the hearing-impaired, promoting appropriate education and training

Solar Ear's founders have observed that many deaf people, especially if they communicate by sign, develop excellent fine-motor dexterity, well-suited for soldering the small circuits of Solar Ear's products.

Solar Ear is radically cheaper than American hearing aids, and the solar battery charger itself, priced at around $50, could provide environmental and economic benefits, slashing battery costs and reducing toxic volume in landfills. One rechargeable battery can replace around fifty disposables, per year. Hearing aid users sometimes have hearing aids in both ears, doubling the savings, and the battery charger lasts two to three years, multiplying the benefits yet more.

YOU

- Help bring Solar Ear to the United States. Its hearing aids and solar-charged batteries, plus charger, would cost much less than available hearing aids. Marketing to middle-income customers could subsidize the provision of hearing aids to low-resource areas.
- When she was just sixteen, Grace O'Brien founded **EarsForYears.org** to bring Solar Ears to deaf children in low-resource regions. Help her grow the effort.

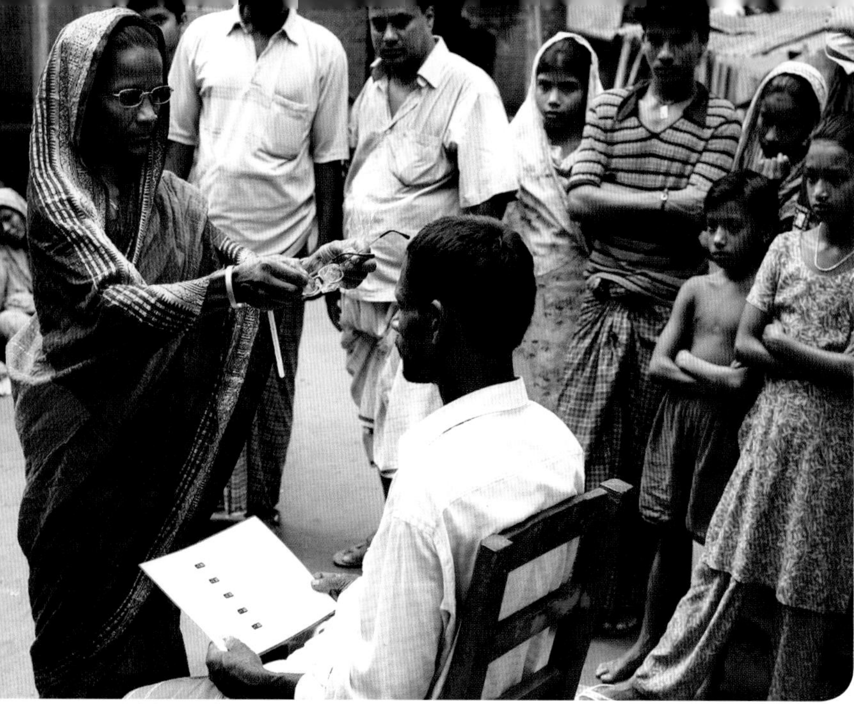

VisionSpring Entrepreneur fits eyeglasses while curious villagers look on, Bangladesh. © John-Michael Maas/VisionSpring

Vision Correction

Hundreds of millions of adults and children in the world lack access to affordable eyeglasses that can boost their literacy and productivity.

 tel

VisionSpring.org • @VisionSpring • VDWOxford.org • @CVDW_

15 When words and numbers blur on the blackboard, texts on the phone are too small to read, or threading a needle is a challenge, the solution is usually simple: eyeglasses. Vision-correcting spectacles have been around for centuries. Manufactured in massive quantities, no-frills glasses are now affordable for global low-income customers—but hundreds of millions of people in the world don't have access to them.

As world populations age, (the average lifespan was seventy years in 2011), coupled with rising literacy rates (**#81**), more people require reading glasses, simple magnifiers. About 10 percent of schoolchildren need prescription glasses. Creative efforts to design and distribute appropriate eyeglasses for low-income customers are expanding to fill this massive need.

Challenges:

- Many people have never had their eyes tested and do not realize they have vision problems or that these problems could so easily be corrected.
- Urban eyeglass establishments are far from rural villages and their prices are unaffordable.
- Vanity is not limited to affluent people; many people are resistant to wearing glasses they perceive as unstylish.
- Children with glasses can be teased.

VisionSpring, a large, well-established social enterprise focusing on reaching customers in low-resource areas, reports that acquiring glasses raises monthly income by about 20 percent, a huge impact for those earning meager incomes. Their successful model has enabled them to sell more than 2 million pairs of attractive yet inexpensive eyeglasses. Their goal is to expand beyond their three focus markets of India, El Salvador, and Bangladesh, to twenty additional countries through new partnerships with social enterprises and NGOs, to reach more of the unmet demand for their high-quality, low-cost, high-impact, attractive eyeglasses.

VisionSpring runs their own urban shops

Twin glasses for Mexican schoolgirls, Oaxaca. © Government of Oaxaca

This owner of OneDollarGlasses sports orange accent beads, Burkino Faso. © Daniel Schiow

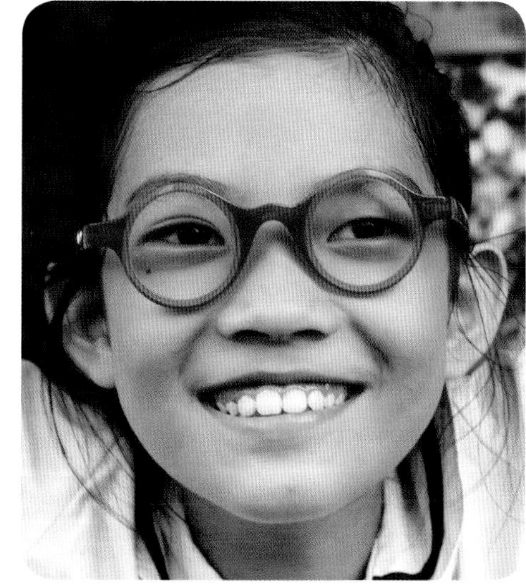

Sporting AdSpecs © Centre for Vision in the Developing World

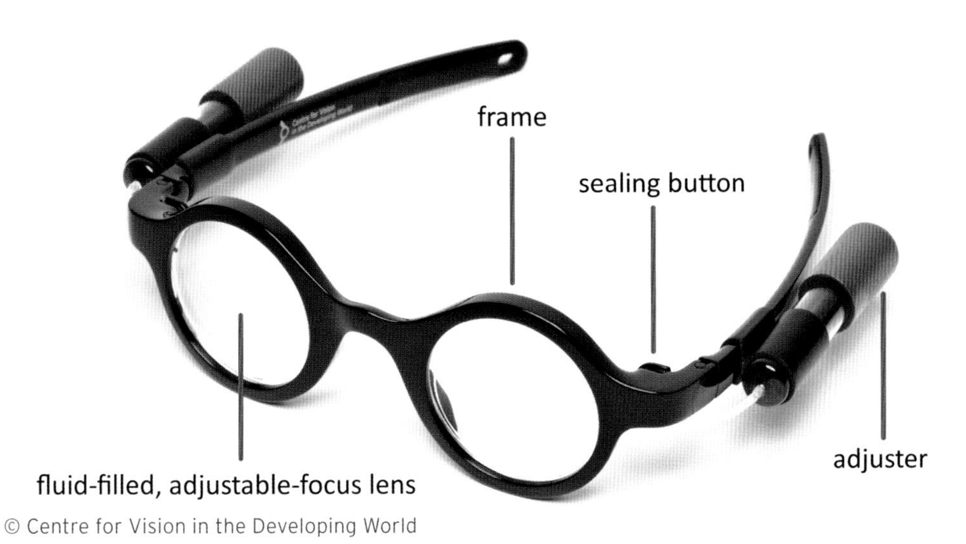

© Centre for Vision in the Developing World

where prescription glasses are made to order. These stores serve as hubs. Vision Entrepreneurs, reaching out to customers far afield, serve as the spokes of their business model. These local representatives, usually women:

- ✔ Travel to rural villages with a tote full of glasses in standard magnifications
- ✔ Test and screen local residents' vision, referring those with serious vision issues to their hub location for prescription glasses or partner hospitals for advanced eye care
- ✔ Educate people about vision issues and the benefits of eyeglasses
- ✔ Provide customer feedback, helping tailor offerings to local customer preference

VisionSpring is donor-subsidized, but working toward sustainability by franchising their eyeglass enterprises to other NGOs, expanding scale and lowering prices. Their cheapest glasses start at $2. They also sell higher-priced glasses aimed at less impoverished customers who are nevertheless priced out of conventional optical shops.

VisionSpring's motto, "We're restoring more than vision, we're helping people see their full potential," attests to the added productivity of corrected sight. After subtracting the cost, the return on customer investment in glasses is $26 for each dollar spent.

See Better to Learn Better/Ver Bien, a governmental partnership with Augen Optical and donors, distributes $10 prescription glasses to Mexican school students, boosting academic performance. An award-winning design permits kids to choose two-tone, snap-together, virtually indestructible glasses. **verbien.org.mx**

OneDollarGlasses, the brainchild of German physics teacher Martin Aufmuth, are made to order from wire and imported lenses. The glasses are fabricated using a toolbox he invented, providing both affordable eyewear and local jobs. Each customer picks colored accent beads. They are so lightweight, nose pads are not needed. For those needing more complex corrections, the self-adjusting **AdSpecs** (short for "adjustable spectacles") are Oxford atomic physicist Joshua Silver's ingenious solution. Requiring no optometrist or optician radically lowers costs and expands accessibility of eyeglasses. A new child-specific design, **Child ViSion**, has won several design awards and is in large-scale trials.

After the correct amount of fluid is injected into the lens, the adjuster tubes are removed. *Voilá!* Inexpensive customized correction eyeglasses. Silver now heads Oxford University's Centre for Vision in the Developing World.

Focus on Vision, a social enterprise in The Netherlands, has created Focusspecs, adjustable lenses regulated by dials. Focus on Vision manufactures their glasses and sells them to NGOs for distribution, emphasizing the importance of end users paying for them (so they value them more than if they had been given free).

> **What About Donating Used Glasses?** A 2011 study published in Optometry and Vision Science reports this approach costs twice as much as providing new eyewear and is not a recommended strategy. Only 7 percent of donated glasses are suitable for reuse.

YOU

- Older women typically buy multiple pairs of readers. Help VisionSpring design and market a product line aimed at this demographic through an online store supporting VisionSpring's expansion.
- Optometrists seeking a continuing education experience in the field should contact VisionSpring.
- **Focusspecs.com** donates one pair of eyeglasses to an NGO partner for each pair sold.

EDU

- Check out the slide show aimed at students posted at **OneDollarGlasses.org**.

Sector 2 – An Overview

GIRLS' AND WOMEN'S HEALTH

In girls' and women's health, there's good news and bad news.

The bad news: Women in low-resource areas routinely die from entirely preventable causes. Their deaths often leave their children motherless, further compounding suffering. Far too many of the world's poorest women have not been reached by the 20th century's extraordinary advances in family planning and reductions in maternal mortality.

"The day of birth is the most dangerous day in the life of a woman and her child. The fact that women do not get the care they need during childbirth is the most brutal expression of discrimination against women. To prevent these tragic and unnecessary deaths is not only a humanitarian urgency of highest priority, but a key investment for social and economic development."

—Former Norwegian Prime Minister, Jens Stoltenberg

The good news: Cheap interventions, like administering a dose of Misoprostol (**#22**) to arrest postpartum hemorrhaging, save women's lives. Before the establishment of the Millennium Development Goals in 2000, which include improving maternal health and lowering maternal mortality, many countries' maternal deaths were not even tracked.

In addition to benefiting from gender-neutral health improvements like those featured in Sector 1, women need gender-specific health services. Low-cost/high-impact approaches are highlighted in this section. Five relate to reproductive health, family planning, and prenatal care.

Expanding reproductive health education is central. One urgent issue for women—and men—is discussing and teaching about the adverse health consequences of female genital mutilation. Education often leads toward communities abandoning the practice (**#16**).

Women who develop cervical cancer, caused by the sexually transmitted human papillomavirus (HPV), typically are in their thirties and forties, so diagnosing and treating this terrible disease (**#17**) not only saves their lives but also protects their families. More good news: There is now an HPV vaccine (**#4**).

Tools from other sectors also impact women's health positively.

Literacy (**#81**) may be the most important poverty alleviation tool of all. Educated women's children are healthier. Educating women correlates with decreased birth rates and improved economic status. Enforcing laws banning young-girl marriages (**#95**) helps keep girls in school and postpones childbearing, improving the health status of adolescent girls and their offspring.

Gender-based violence extracts an enormous health toll on girls and women. Passage of the **International Violence Against Women Act (IVAWA)** (**#96**) would help the United States prioritize global efforts for women's freedom from this ever-present threat.

Other interventions help, too, sometimes in surprising ways. Some examples include:

- ✔ Sanitary napkins for schoolgirls (**#47**) minimize their exposure to attack and ridicule and improve the likelihood they will stay in school.
- ✔ Carrying portable solar lights (**#27**) to the latrine at night lessens women's vulnerability to attack.
- ✔ Providing bikes (**#78**) to schoolgirls not only saves them, time it also decreases pregnancy rates because they are not dependent on older men for rides.

It is simple: Healthier girls and women have a better shot at moving themselves and their families up out of extreme poverty.

16. REPLACING FEMALE GENITAL MUTILATION WITH ALTERNATIVE RITES OF PASSAGE

Eradicating FGM will prevent millions of girls from suffering bodily harm; alternative ceremonies ease cultural acceptance of abandoning this practice.

17. CERVICAL CANCER VINEGAR SCREENING

Cervical cancer, the leading cause of women's cancer deaths in the Global South, can be detected with a simple vinegar application and treated during the same visit.

18. CONTRACEPTION

Jadelle, two tiny inserted rods, provides contraception for five years and is completely reversible. Plan B prevents pregnancy from unprotected sex.

19. PRENATAL AND POSTNATAL MATERNAL NUTRITION

Better prenatal and postnatal nutrition improves outcomes for both mother and baby.

20. CLEAN BIRTH KITS

Clean Birth Kits provide sterile, disposable tools for safe birthing, protecting both mothers and newborns from potentially deadly infections.

21. NONSURGICAL MEDICATION ABORTIONS

Nonsurgical medication abortion (accomplished via pills) is a safe, noninvasive procedure.

22. MISOPROSTOL: SAVING MOTHERS' LIVES

Misoprostol, an inexpensive generic drug, prevents and treats post-partum hemorrhage, the cause of the majority of maternal deaths worldwide.

Nelly, Soine, and Tayaina share a quiet moment during their Alternative Rite of Passage, Kenya. © Teri Gabrielsen/African Schools of Kenya

Replacing FGM with Alternative Rites of Passage

Eradicating FGM will prevent millions of girls from suffering bodily harm; alternative ceremonies ease cultural acceptance of abandoning this practice.

tostan.org • @tostan • ASKenya.org • @ASKenya • EdnaHospital.org • @EdnaAdan

The movement to end **female genital mutilation (FGM)** is gaining momentum; numbers are declining. This ancient social norm is still widely practiced in Africa and in some areas of Asia and the Middle East. It is performed on young girls, without their consent, in an excruciatingly painful and traumatic procedure.

FGM can cause life-threatening complications and harmful long-term effects including:

- ✗ Infection
- ✗ Hemorrhage
- ✗ Shock
- ✗ Chronic pain
- ✗ Urinary tract infections
- ✗ Painful urination
- ✗ Menstrual problems
- ✗ Infertility
- ✗ Increased labor complications and newborn deaths
- ✗ Fistulas following childbirth

Stigmas are attached to these side effects, literally adding insult to injury. Girls occasionally die from the procedure.

> "FGM has no health benefits, and it harms girls and women in many ways. It involves removing and damaging healthy and normal female genital tissue, and interferes with the natural functions of girls' and women's bodies."
>
> —*World Health Organization*

The 2012 United Nation's resolution titled "Intensifying Global Efforts for the Elimination of Female Genital Mutilations" calls on all states to enact legislation banning FGM. African women have led the initiative, denouncing FGM as:

- ✗ A human rights violation
- ✗ A form of violence against women (**#96**)

Important as laws are, they have not resulted in ending FGM. Enforcement is virtually nonexistent. Educating communities about misconceptions regarding FGM is central to ending the practice, since elders and mothers perpetuate the tradition.

- FGM *causes* health difficulties, a fact that many women do not know. In regions where nearly all girls are cut—in Somalia, for example, 98 percent—women assume the health problems caused by cutting are naturally occurring phenomenon. Learning that they're entirely avoidable helps in reevaluating the tradition.
- Islam does not sanction cutting; many women incorrectly assume it is a religious requirement. The support of religious leaders—promoting its abandonment as a way to improve girls' and women's health—is crucial to FGM's eradication.
- Men do not necessarily support or expect cutting. Including men in the discussion has been a breakthrough. Cutting has been a private women's tradition about which men have little knowledge. Knowing men do not require FGM builds mothers' confidence that uncut daughters will have suitable marriage prospects. Many men are speaking out in favor of eradicating FGM.

Tostan, a Senegal-based NGO, is a leader in the movement to abandon cutting. Tostan's terminology, **female genital cutting (FGC)**, is used in place of female genital mutilation. Their community workers find less judgmental, value-laden language more effective in facilitating change.

In 1997, women participants in Tostan's literacy/health module in the village of Maicounda Bambara came to the independent, courageous decision to stop cutting their daughters. In lieu of the tribal cutting celebrations, they staged a public **Declaration Event** ritual announcing this path-breaking decision. Other villages followed suit, and leaders realized replacing the old rite with a new celebration was helpful in easing people to accept FGC's abandonment.

A youth theater group bids "Adieu" to female genital cutting, Ziguinchor, Senegal. © Tostan

Multiple villages now join together to celebrate their abandonment of FGC in public "Declaring" festivities, consolidating this change. Men join in, important for communal solidarity. More than 6,500 communities from Djibouti, Guinea, Guinea-Bissau, Mali, Mauritania, Senegal, Somalia, and The Gambia have declared their abandonment of FGC through traditional oratory, dance, music, and storytelling.

Tostan is growing and expanding its influence in Western Africa. Pledges to end FGC are now frequently paired with commitments to end child marriages (**#95**).

Alternative Rites of Passage were recently introduced in Kenya for 52 young Maasai girls, including a chief's daughter, a cutting-free initiation into womanhood. Celebrating their commitment to ending FGM at a tribal gathering both honors tradition and creates a new rite of passage for their daughters. It is a festive, dramatic moment for the whole community. Curriculum taught by elders includes:

- Human growth and development
- FGM—myths, misconceptions, and health risks
- Family planning
- HIV and STD prevention
- Human rights and self-esteem

The community realized the importance of providing elder women cutters not just respect, but new livelihoods going forward. During the ceremonies, goats (**#72**) were gifted to each of four former cutters, along with community recognition and admiration for their willingness to renounce cutting.

Edna Aden Ismail tirelessly advocates for the eradication of FGM. Her capstone achievement, **The Edna Aden University Hospital,** is more than 4,000 miles east of Senegal on Africa's east coast in Hargeisa, Somaliland. The hospital's primary mission is "to improve maternal and infant health care, and fight the practice of female genital mutilation."

The hospital treats women suffering from long-term, debilitating effects caused by cutting, making the staff especially passionate about ending it forever. Their research department collects data on the prevalence and harms that FGM causes, helping target eradication strategies.

YOU

- Edna welcomes volunteers with specialized skills:
  - health practitioners, or students of nursing, midwifery, medicine, or biochemistry
  - teachers of English as a foreign language, for staff and students
  - library helpers and book cataloguers; computer and IT support staff
  - statistics and data collection experts to assist with ongoing research on FGM

- Tostan runs a one-year volunteer program. College students and recent grads, as well as master's and doctoral candidates, with French proficiency are especially encouraged to apply.

Nurse Marilyn Isaacs Mingo prepares for a cervical cancer screening, Guyana. © Maureen Reinsel, Courtesy of Jhpiego

Cervical Cancer Vinegar Screening

Cervical cancer, the leading cancer killer of women in the Global South, can be detected with a vinegar application and treated during the same visit, saving many women's lives.

jhpiego.org • @jhpiego • pincc.org • @PINCClink • @GroundsForHealth.org • @grounds4health

Nurse Marilyn Isaacs Mingo is not mixing up dinner. She is prepping for her day's cervical cancer screenings, pouring acetic acid. That's vinegar. This low-tech procedure prevents many cervical cancer deaths.

- Cervical cancer rates in affluent countries have been slashed, due to generations of Pap smears.
- 273,000 women die annually of cervical cancer worldwide, overwhelmingly in the Global South.
- Almost all cervical cancers are caused by the human papillomavirus, a sexually transmitted disease (STD).
- HIV-positive women have a higher risk of HPV infection and therefore also are at higher risk for contracting cervical cancer.
- Cervical cancer typically kills women in their thirties and forties, often leaving young children motherless.
- The newly introduced HPV vaccine (**#4**) prevents most cervical cancer.

This low-cost vinegar-testing procedure was developed by **Jhpiego**, an affiliate of Johns Hopkins Medical School. The vinegar solution is applied to the cervix. Precancerous lesions turn white within a few minutes, giving the treatment its name: **Visual Inspection with Acetic Acid (VIA)**. Lesions are frozen with carbon dioxide or nitrous oxide gas using a sterilized metal rod at the same visit: Screen-and-Treat. VIA and cryotherapy together is called **VIA/Cryo**.

- VIA requires no lab work.
- VIA/Cryo workers can be trained in a few weeks.
- Freezing is about 90 percent effective, causing a burning sensation that fades in a day or two.
- There is no long wait for test results, nor is any return visit needed.
- If advanced cervical cancer is discovered, surgery is required.

The **Cervical Cancer Prevention Program in Zambia (CCPPZ)** runs seventeen clinics. When women are screened, digital photos are sent for remote examination, using **Electronic Cervical Cancer Control (EC3)**. In 2013, they completed their 100,000th screening. They also train health care providers from other countries. **acewcc.org**

> If clinics don't own freezing equipment, they often borrow or rent a carbon dioxide tank from the local Coca-Cola bottling plant.

Prevention International: No Cervical Cancer (PINCC) sends teams to India, Latin America, and East Africa, providing training and helping establish VIA/Cryo programs.

Grounds for Health (GFH) has established VIA/Cryo programs in coffee-growing communities in Mexico, Nicaragua, Peru, and Tanzania. When coffee cooperatives request help, GFH trains volunteer community health promoters to educate members about cervical cancer and screening. They then instruct local health workers and provide transportation for women to travel to the screening site.

- **PINCC** actively recruits both medical professional and lay volunteers.
- **GFH** welcomes volunteer nurses, nurse practitioners, midwives, and physicians.
- **GFH** is part of Ebay's Giving Works. 10 percent–100 percent of sales can be allocated to Grounds for Health programs.

Contraception

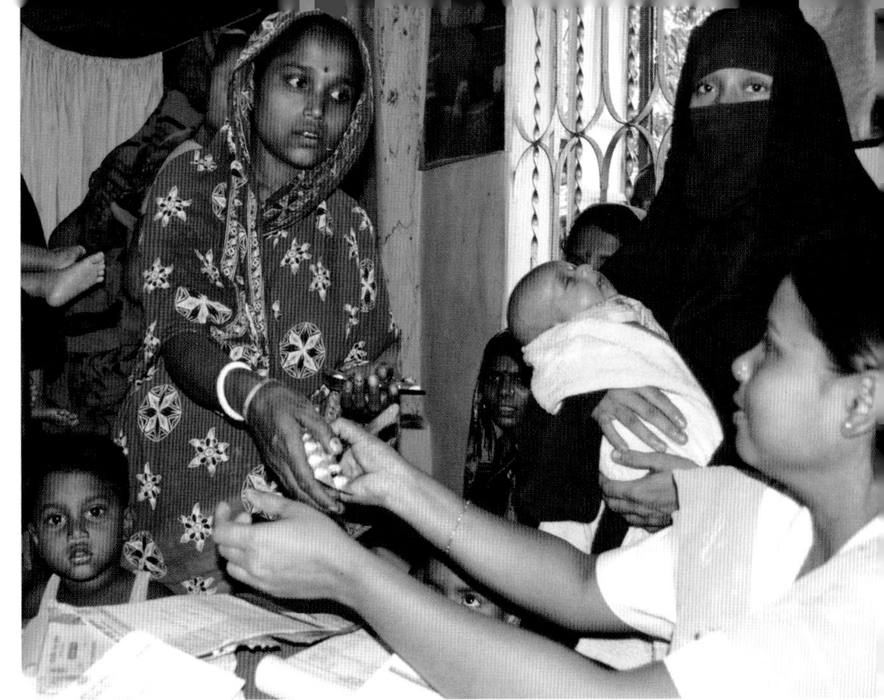

Demand for family planning in Bangladesh. Can you find the five children in the photo? © T. Cody Swift/ Pathfinder International

Jadelle®, two tiny inserted rods, provides contraception for five years and is completely reversible. Plan B prevents pregnancy from unprotected sex.

ippf.org • @ippf • mariestopes.org.uk • @MarieStopes • pathfinder.org • @PathfinderInt

Worldwide, more than 200 million women desiring to limit their fertility lack modern contraception. A smorgasbord of birth control options exist—for women lucky enough to access them. Contemporary contraceptives have fewer side effects, are more effective, and leapfrog over previous awkward, inconvenient, and indiscreet methods.

Family planning services are labor intensive. Most contraception can't just be handed out assembly line–style like deworming medications. **Jadelle**, approved by the World Health Organization, was developed by The Population Council; one implant lasts five years. (Nexplanon®, a similar product, is used in the United States.) It has the potential to fill some of the vast unmet need for modern contraception. Jadelle is:

- ✔ Affordable: In 2012, through **The Jadelle Access Program**, the price per implant dropped to $8.50 for low-income countries in a complex deal negotiated by multiple players.
- ✔ Convenient: Poor women in remote areas, the most underserved, are spared the arduous and costly challenge of frequent clinic visits.
- ✔ A boon for fragile healthcare systems: Health workers spend less time dispensing contraception, since Jadelle lasts so long; their time is freed up for other healthcare activities.

- About the size of two match sticks, Jadelle is inserted in the inside upper arm, taking just a few minutes under local anesthesia. It is removed the same way.
- It lasts five years; upon removal, fertility resumes within a few days.
- Its main side effect is changes in menstrual bleeding patterns. Some women have irregular bleeding; others experience no bleeding.
- Jadelle is as effective at preventing pregnancy as sterilization, but is reversible.
- The main side effect of leaving it in beyond five years is pregnancy.

Jadelle does not protect against STDs.

> "When fully implemented, the Jadelle Access Program will avert more than 28 million unintended pregnancies between 2013 and 2018, and, ultimately, prevent approximately 280,000 infant and 30,000 maternal deaths."
>
> —*Bill & Melinda Gates Foundation*

Emergency contraception is an important tool for women following forced and/or unprotected sex. Nicknamed "morning after pills," Plan B and similar brands prevent ovulation.

With rape being a frequent tool of war and conflict, and gender-based violence rampant worldwide (**#96**), emergency contraception is crucial for victims, allowing them to prevent violence-conceived pregnancies. (Post-exposure prophylaxis to prevent the transmission of HIV should also be provided.)

Emergency contraception is available over the counter in many countries, including the United States.

A reproductive health class. IUDs are the most popular choice in Nigeria. © Marie Stopes International/Glenna Gordon

"Every mother at our mobile clinic gets prenatal vitamins each month from Vitamin Angels. At one site, we saw 100 women! You can do the math, but that's 3,000 vitamins in one day!" © Cheryl Hanna-Truscott/Midwives For Haiti

Prenatal and Postnatal Maternal Nutrition

Better prenatal and postnatal nutrition improves outcomes for both mother and baby; distribution is the challenge.

VitaminAngels.org • @VitaminAngels • sghi.org

19 Prenatal and postnatal care is not a single item but rather a basketful of interventions to help women give birth to and raise the healthiest babies possible while remaining healthy themselves.

One of the biggest dangers for pregnant women in low-resource regions is calcium, iodine, and iron deficiency (**#7**). Women need additional iron when they are pregnant. WHO estimates 20 percent of maternal deaths involve mothers with anemia, insufficient iron resulting from:

- General malnourishment
- Urban diets with less access to green, leafy iron-rich vegetables
- Vegetarian diets, inherently low in iron
- Menstruation

When Jennifer Tsai and Matthew Edmundson were MBA students at New York University's Stern School of Business, they teamed up to design a social enterprise to deliver iron to pregnant women in India, with a 60 percent rate of prenatal anemia, one of the world's highest. Their market research revealed that among women with access to iron pills, compliance was very low. They disliked the aftertaste and unpleasant side effects.

Deeper resistance came from perceiving pills as medicines rather than beneficial supplements; since they were pregnant, not ill, they didn't see any reason to take them.

Tsai and Edmundson's solution—an NYU competition winner—is **Violet Health**, a company marketing iron-fortified biscuits. These cookies will be a for-profit product in upscale markets and subsidized in low-income communities. Indian women, like most people, prefer cookies to capsules.

Several other prenatal and postnatal supplements are available. **Sprinkles™**, a children's micronutrient powder packet, is now available formulated for pregnant and lactating women, addressing iron and calcium deficiency.

Sprinkles™ micronutrient packets. © www.sghi.org

Plumpy'Mum™ is new from Nutriset, the manufacturer of Plumpy'Nut™ (**#12**). Designed for both pregnant women and lactating mothers, it addresses the maternal nutritional deficits that often result in vulnerable, low birth weight babies, and fosters postnatal growth.

Raj-Nutrimix (**#7**) is a powdered micronutrient supplement packaged in Banswara, India by the local Women's Self Help Group in partnership with the **World Food Programme** and the **Global Alliance for Improved Nutrition (GAIN)**. Formatted for pregnant women, lactating mothers, infants and young children, it is mixture of fortified wheat, soy, maize, sugar, twelve vitamins, and six minerals.

Biofortified crops (**#59**) approach the problem differently. While consumed like any other food, they deliver more iron. Pearl millet is naturally higher in iron than the more common wheat, rice, and maize. HarvestPlus is developing new, drought-resistant, higher-iron millet named *dhanshakti*, meaning "prosperity and strength."

YOU

- **VioletHealth.com** is a for-profit social enterprise. Tsai and Edmundson are seeking Angel Impact Investors.

Clean Birth Kits

Clean Birth Kits provide sterile, disposable tools for safe birthing, protecting mothers and newborns from deadly infections.

tel

CleanBirth.org • @CleanBirth • ayzh.com • ayzhInc

An ayzh assembly line in Chennai, India, packaging Clean Birth Kits purses. © ayzh/Justin George

Ziploc bags of inexpensive sundries don't look like life-saving equipment, but their power to prevent infections saves mothers' and newborns' lives. When mothers die giving birth, their newborns' survival odds plummet; maternal and newborn mortality are inextricably linked. Motherless older children are at higher risk, too.

Clean birth kits have been around for decades. Worldwide about 60 million women annually give birth at home, often on dirt floors, without clean water. Infections cause about 10 percent of the 360,000 annual maternal deaths.

Kits provide tools for the "Six Essential Cleans" outlined by WHO:

- ✔ Clean hands—soap or sanitizing wipes
- ✔ Clean perineum (birth canal)—gauze wipes
- ✔ Clean birthing surface—folded plastic drop cloth or padded, blood-absorbent sheet
- ✔ Clean cord cutter—sterile razor blade or scalpel
- ✔ Clean cord-tying thread
- ✔ Clean cord care—could be CHX ointment (**#5**)

Kits include pictorial directions. Upgrades include plastic gloves, sanitary pads for post-partum use, towels, and newborn blankets and/or clothing.

> **Low-income countries' hospitals frequently run out of sterile supplies – women are instructed to bring their own.**

While giving birth in an Indian hospital, Zubaida Bai contracted an infection that caused her years of suffering. She eventually learned she would be unable to have more children. A mechanical engineering and design professional, Bai put her passion and skills to work providing women access to safe birth so no one else would ever have to live through her experience.

The result, **azyh**, (pronounced "eyes") is a social enterprise producing a stylish $2 Clean Birth Kit dubbed the *JANMA*, meaning "birth" in Hindi. Attractive reusable jute purses replace Ziplocs, with clean birth tools tucked inside. The kits are assembled in India, and the company provides local women jobs. Their customers are both

The ayzh safe birth kit in a purse. © azyh

nonprofits and commercial resellers, a "business-to-business" model. More offerings are in the pipeline, packaged in their signature bags. They are building a trusted, affordable, well-designed women's health tools brand.

Australian blogger Adriel Booker started **Bloggers for Birth Kits**. Following her lead, bloggers have assembled more than 7,000 kits, transported by ship to Papua New Guinea, where very few mothers give birth in safe conditions. @AdrielBooker

YOU

- **CleanBirth.org** works in Laos, where maternal and infant mortality rates are among the world's highest. They offer baby shower favors, each card representing the donation of an **ayzh** Clean Birth Kit.

YOU EDU

- United Methodist Committee on Relief collects Clean Birth Kits packaged by volunteers. Directions for assembling and shipping are on their website. Kits include baby blankets and onesies, and are sent to areas with the greatest humanitarian need. **www.umcor.org**

2013 Hot Line Kick-Off in Bangladesh. © Asia Safe Abortion Partnership

Nonsurgical Medication Abortions

Nonsurgical Medication Abortion (accomplished via pills) is a safe, noninvasive procedure.

WomenOnWaves.org • @WomenOnWeb.org • @abortionpil • @Ipas.org • @IpasOrg

Abortions induced by medications, termed **medical abortions**, are noninvasive, nonsurgical procedures for ending early pregnancies up to nine weeks after a woman's last menstrual period. They involve two medications:

Mifepristone (RU-486), which blocks the hormone progesterone. Without progesterone the attachment of a fertilized egg to the uterine wall is disrupted.

Misoprostol, also used to treat postpartum hemorrhage (**#22**), causes uterine contractions. (Misoprostol can be used by itself, but without Mifepristone, this is less foolproof.) Miso, originally an ulcer medication, has been used for safe medicinal abortions since the mid-1980s.

Many organizations work to make medical abortions, and knowledge of administering them, available to women in low-resource areas. Without access to safe procedures, women are forced to resort to unsafe options.

WomenOnWaves, founded by Dutch physician and human rights activist Rebecca Gomperts, sails its ship near countries where abortion is illegal, treating women and highlighting the right to safe abortion. Its sister organization, **WomenOnWeb**, provides detailed, medically accurate information about Mifepristone-Misoprostol abortion procedures, including downloadable instructions in eight languages.

They partner with in-country initiatives, launching hotlines like the one in Bangladesh (pictured). **Línea Aborto Información Segura** has provided information to thousands of women in Chile, where abortion laws are among the most restrictive in Latin America.

North Carolina–based **Ipas.org** advocates for women's access to safe abortions—via pills or other safe techniques—partnering with organizations in more than twenty countries.

> "No woman should have to risk her life, her health, her fertility, her well-being, or the well-being of her family . . . We struggle against the fundamental social injustice resulting in the deaths of so many women in the prime of their lives."
>
> —*IPAS*

When abortion access is restricted, women resort to unsafe abortions, with frequently dangerous outcomes, including death. WHO defines an unsafe abortion as "the termination of a pregnancy by those lacking the necessary skills, or in an environment lacking minimal medical standards, or both."

Unsafe abortions kill many women and put the lives of countless children left motherless at higher risk, too.

- ✗ 47,000 women die from complications of unsafe abortions each year; that's 13 percent of all maternal deaths.
- ✗ 85 percent of deaths from unsafe abortions are in the developing world.
- ✗ Abortion rates are not lower in countries where it is restricted. Women risk their lives to end unintended pregnancies.
- ✗ Rape victims who get pregnant and are denied safe abortions are doubly victimized.
- ✗ Infections and perforated uteruses are the most frequent causes of death from unsafe abortion.

In the United States, legally induced abortion results in 0.6 deaths per 100,000 procedures. Worldwide, unsafe abortion accounts for a death rate that is 350 times higher (220 per 100,000), and in Sub-Saharan Africa the rate is 800 times higher, at 460 per 100,000.

—Guttmacher Institute

Misoprostol: Saving Mothers' Lives

Misoprostol, an inexpensive generic drug, prevents and treats post-partum hemorrhage, the cause of the majority of maternal deaths worldwide.

LifeforAfricanMothers.org • @Life4AM • VSInnovations.org • VSInnovations

Newly graduated Skilled Birth Attendants Micheline Leurbour, Carline Jean-Gilles, Ysemonique Thelesmon, and Fedeline Magmy will improve maternal outcomes, St. Therese Hospital, Hinche, Haiti. © Cheryl Hanna-Truscott/Midwives For Haiti

All pregnant women need extra iron. Undernourished women are especially susceptible to iron deficiency anemia (**#19**) making them more likely to die from postpartum hemorrhage, excessive bleeding following childbirth. It also puts them at higher risk of death from infection.

- ✗ 99 percent of worldwide maternal deaths are in the developing world. The majority of these deaths are due to postpartum hemorrhage.
- ✗ Maternal death rates are higher for women in rural areas where skilled health care providers and facilities are sparse and transportation difficult.

Misoprostol, an inexpensive pill requiring no refrigeration, is administered orally to prevent or treat postpartum hemorrhage. "Miso," under a dollar a dose, stimulates uterine contractions, stemming excessive postpartum bleeding.

Misoprostol joined WHO's List of Essential Medicines in 2012. Uptake varies by country. NGOs, intergovernmental health organizations, and health ministries partner to create Miso supply chains.

Misoprostol also is used for safe medical abortions (**#21**) and for treatment of incomplete abortions, saving women's lives. In countries where abortion is outlawed or heavily restricted, obtaining Misoprostol—even to treat life-threatening postpartum hemorrhaging—can be difficult, resulting in preventable maternal deaths.

The goal is for women to give birth in facilities with trained attendants and supplies. For many low-income, rural women, though, home births are the norm. With luck, a Skilled Birth Attendant arrives in time. Optimally, Skilled Birth Attendants develop advance birth plans with mothers-to-be, including instructions and Misoprostol doses for unattended home births.

Life for African Mothers (LFAM), which operates in eleven countries, is "a maternal charity with a simple mission: providing cheap medications that save women's lives."

Angela Gorman, a retired Welsh neonatal nurse, was mobilized to action after watching the 2005 BBC documentary *Dead Mums Don't Cry*, chronicling maternal deaths in Chad. She unretired and set to work building LFAM into an effective organization, saving thousands of women's lives by supplying meds, promoting training, and advocating for prenatal care in Sub-Saharan Africa.

Venture Strategies Innovations launched in 2008 to assure, given Misoprostol's low cost and ease of use, that women who need it most have access. They target countries with high maternal death rates, and now work with more than twenty Asian and African countries, promoting distribution of Misoprostol (at $1 a dose) and other life-saving interventions.

YOU

- Misoprostol is well-suited for marketing promotions. Retailers: Donate one dose per sale.
- Healthcare professionals can volunteer with **LFAM**. "Short projects with LFAM can make a massive impact in the communities where we work, but professionals who work with us often talk of how beneficial the experience has been for them personally and professionally."
- **MidwivesforHaiti.org** also welcomes medical professional volunteers.

Sector 3 – An Overview

LIGHTS, CELL PHONE CHARGING, ENERGY!

The industrialized world generates most of its electricity in costly, large, centralized plants typically fueled by coal, natural gas, hydroelectric power, or a nuclear reactor. Power lines transmit electricity directly to customers' homes. We press buttons, flick switches, plug into outlets, and things run. Users rarely notice electricity except during rare, newsworthy power outages.

Electrification took off in the early 20th century, primarily providing lighting. A constant stream of life-enhancing, labor-saving devices has followed. In low-income countries, an estimated 1.3 billion people are not connected to power grids. The benefits of domestic electricity have bypassed them.

Many rural villages remain largely non-electrified, as do ever-expanding informal settlements on the periphery of cities. There is insufficient capacity for constructing centralized power plants and running transmission lines to remote areas or to urban slums lacking infrastructure.

Energy poverty—lacking access to affordable power—forces people to patch together the little power they can afford purchasing batteries, kerosene, wood, dung, crop residue, and charcoal.

These energy sources:

✗ Cost more than grid electricity
✗ Emit smoke, endangering health
✗ Cause fires and burns
✗ Pollute the earth through black carbon emissions and unsafe battery disposal
✗ Cause extensive deforestation

Rather than waiting for the grid to arrive, on-site energy generation products—called **distributed energy**—are becoming affordable alternatives. Dropping solar technology prices could potentially reduce energy poverty by displacing inferior kerosene lighting.

Cell phones, entry-level workhorses of the developing world (**#84**), require frequent charging, radically increasing electrical demand among off-grid cell phone owners. Many new solar lamp designs incorporate cell phone chargers. In some cases the demand for phone charging is actually driving the upsurge of solar lighting.

Supplying energy products to off-grid customers is both an enormous challenge and a gigantic business opportunity. Educating consumers and creating supply and distribution chains for new products requires ingenuity and nimble business models.

Solar-powered lanterns (**#27**) are a short-term approach. Solar systems powered by rooftop panels (**#26**) can electrify many more appliances, and as money permits, more panels can be installed.

Batteries remain a central piece of the energy puzzle. Innovative schemes such as battery recharging (**#23**) yield profits for businesses while also supplying customers more energy for less money. Harnessing human power leveraged by bikes is another promising off-grid power solution (**#30**).

Green Fuel is part of the energy equation as well. Cleaner, more efficient cooking fuels have multiple benefits (**#32, #33**) and align with improved cookstoves (**#50**).

Labor-saving devices spare women toil and time. Additional energy can power grinders, mills, cookers, and sewing machines.

Women and children derive enormous health benefits from eradicating kerosene lamps and open-fire cooking. Eliminating smoke inhalation radically decreases eye and respiratory disease; cooking more efficiently uses less fuel, lightening wood-carrying burdens.

Clean energy decreases women's time spent laundering and scrubbing. The same smoke blackening lungs and polluting the Earth's atmosphere settles as soot on surfaces and clothes.

The following entries show innovative, effective approaches to at long last em**POWER** the world's poorest women.

23. CENTRALIZED BATTERY RECHARGING

Single-use batteries are the costliest form of energy. A battery recharging business delivers higher-quality, lower-cost, eco-friendly batteries.

24. BRIGHT BOX SOLAR-CHARGED BATTERY

The Bright Box, charged by a small solar panel, powers two lights and a radio, and also charges cell phones and other electronic devices.

25. LED BULBS

LED lightbulbs provide affordable, ultra-efficient, high-quality illumination and the potential to light up off-grid communities worldwide.

26. SOLAR HOME SYSTEMS

Rooftop solar panels provide residential power to turn on the lights, charge mobile phones, and run small TVs.

27. PORTABLE SOLAR LANTERNS AND LAMPS

Portable solar LED lamps and lanterns are increasingly affordable replacements for kerosene. Lamps provide superior light while eliminating health hazards and fuel costs.

28. SUNSALUTER

The SunSaluter facilitates a solar panel's rotation from east to west, generating 40 percent more solarization than if the panel is simply lying flat.

29. PAY-TILL-YOU-OWN ELECTRICITY

Solar panels for off-grid, nonelectrified households can be installed and paid off in weekly installments via mobile phone.

30. BIKE-POWERED MACHINES

Bicycles can leverage human power to run a wide variety of specially engineered machines, performing tasks quickly without consuming electricity or fuel.

31. BIOGAS BACKPACK

Biogas backpacks provide a low-tech method for biogas, produced in a central biodigester, to be transported for use as cooking fuel.

32. ECO-BRIQUETTES

Eco-briquettes—fuel fabricated from waste products—burn cleaner, create local jobs, and preserve forests.

33. GREEN CHARCOAL: BIOCHAR BRIQUETTES

Biochar-based briquette production utilizes crop waste to produce high quality, inexpensive cooking fuel and fertilizer while decreasing deforestation.

Eva Nartey, multitasking with a Burro portable phone charger powered by four rechargeable AA batteries, juices her phone while she uses it, Bomase, Ghana. © Steven Adusei/BurroBrand

Centralized Battery Recharging

Single-use batteries are the costliest form of energy. A battery recharging business delivers higher-quality, lower-cost, eco-friendly batteries.

BurroBrand.biz • @BurroBrand • Appropedia.org • @appropedia

Batteries are vital for people not hooked up to the electrical grid, used chiefly for powering flashlights and radios. Though recharging batteries is a more economical approach, therein lies the problem—how do you power the battery charger? Reliance on single-use batteries, far more expensive than grid energy per kWh, is a poverty trap. More than 1.3 billion people lack grid connection; this is often termed **energy poverty**.

The enormous popularity and wide penetration of cell phones with their daily charging needs has increased this energy deficit, sending hundreds of millions of people off on frustrating daily quests for cell phone charging stations.

BurroBrand is an innovative social enterprise, launched in 2009 in Ghana, a country whose population is 84 percent off-grid. Burro's solution is rechargeable battery rental services, including a battery-operated cell phone charger.

- ✓ Clients pay up-front deposits for weekly battery delivery.
- ✓ "Fallen" batteries are picked up for recharging and fresh batteries are delivered.
- ✓ Customers' costs are lower than when they purchase single-use batteries.
- ✓ Their batteries are superior to the imported carbon-zinc cells they replace.
- ✓ Mother Earth benefits from thousands fewer batteries manufactured and trashed.

Burro was founded by an American, Whit Alexander, whose roots go back to his early career in international development in West Africa. He concluded, as have many others, that poverty cannot be alleviated by handouts or the building of large projects whose benefits rarely reach low-income villagers. Some of Burro's clientele live in villages with power lines running overhead without a single household having ever been connected; they are literally disempowered.

Smart, innovative local businesses providing jobs, well-designed products, and excellent customer service help customers climb up the economic ladder. Local businesses can tap into employees' entrepreneurial talents and develop local solutions for local problems.

Burro has added battery-operated cell phone chargers using four AA batteries (pictured above) prototyped by a Burro employee. He developed the product after observing the hoops people jumped through for mobile charging. Handing off their dead phone to a traveling vendor or traveling to far-off charging locations themselves was disruptive, frustrating, costly, and time-consuming. Burro's home-based battery-powered chargers save customers time and money and provide consistency: no more closing down stores to go charge phones. Burro designs its products, which are then manufactured in China (as are the disposable batteries they displace).

Burro is also growing a network of recharging stations providing a charging at a fraction of the cost of buying single-use batteries.

YOU

- Intern with Burro: "Each summer, we pair the best and brightest interns from Ghanaian universities with those from abroad to create high performance teams."
- Consider Burro for an early-stage **impact investment**.

Open-Source Bright Box Batteries

The Bright Box, charged by a small solar panel, powers two lights and a radio, and also charges cell phones and other electronic devices.

OneDegreeSolar.com • @onedegreesolar

A future customer shows off One Degree's signature Solar Orange Box charger, Liberia. © Gaurav Manchanda/One Degree Solar

Solar power has the potential to light up low-income, off-grid households, but it is a hard market to penetrate. Customers are very poor; even if buying rechargeable devices saves them money in the long run, they need to be convinced performance is as promised. Customers also need credit, allowing them to pay with savings accrued from eliminating their kerosene, battery, and cell phone-charging outlays.

For a company to succeed in low-resource regions, it must not only attract customers but simultaneously manage:

- Supply chains
- Tariffs
- Inventory
- Quality control
- Employees
- Customer relations
- Credit schemes to help customers purchase products

The profit margins are very small, making a large customer base essential. More and more companies are bringing low-cost products to the market. Africa alone has more than forty solar companies/products.

One Degree Solar, based in Kenya and expanding across West Africa, Southeast Asia, and South America, manufactures and sells their **Bright Box** based on **open-source** hardware. Designed by CEO Gaurav Manchanda and priced at $80, it can be used to power many different devices. The system:

- ✔ Powers up via a portable 5.4-watt solar panel
- ✔ Includes two bulbs that plug directly into the charged battery and attach to long cords. Owners hang them where they need light. Two more lights can be powered by the box.
- ✔ Charges phones, smart phones, and tablets
- ✔ Plays a radio
- ✔ Meets **Lighting Africa**'s recommended performance targets
- ✔ Pays for itself within a few months and then provides no-cost energy

Urban customers who have electrical connections and more sophisticated devices purchase them for backup power charging when the grid fails, which occurs frequently in the developing world. Business owners like that they expand store hours and reduce lighting expenses.

One Degree Solar has partnered with **Kiva.org**, a microfinance site where people lend money at no interest to help expand opportunity for off-grid people, in this case helping them to access distributed power.

One Degree Solar is a **triple bottom line** social enterprise achieving social, financial, and environmental returns: "profit, people, and planet."

Eesha Khare of Saratoga, California, with her Intel Award. © Chris Ayers/Intel

Instant cell phone charging? Khare took top honors at the 2013 Intel International Science and Engineering Fair, beating out 1,600 finalists. She designed and demonstrated her tiny super-capacitor's ability to power a device in thirty seconds.

YOU

- Check the **Kiva.org** website. One Degree Solar loans are occasionally available.
- One Degree Solar welcomes more distributors.

An admirer of new Quetsol LED lights, turned on via wind power, replacing kerosene lanterns, Guatemala. © Juan Rodriguez

LED Bulbs

LED lightbulbs provide affordable, ultra-efficient, high-quality illumination and the potential to light up off-grid communities worldwide.

LUTW.org • @LUTWgroup • lightingafrica.org • @LightingAfrica

25

LEDs (Light-Emitting Diodes) outperform other types of electric light bulbs, and all electric lights are far superior to kerosene lamps and candles. LEDs are costly up front but they pay for themselves by consuming very little energy and lasting many years. Prices are dropping as LEDs become mainstream. Performance monitoring is being implemented to guarantee high-quality products.

LEDs have been around since the 1960s, but it was Dave Irvine-Halliday, a professor of mechanical engineering from Calgary, Alberta, who first grasped their potential as a path out of **energy poverty** for people lacking grid electricity. While on sabbatical in Nepal, he experienced rural, non-electrified life firsthand and committed himself to tackling this inequity.

Back in Calgary, he and an assistant checked out a white-light-emitting diode produced by Nichia, a Japanese company. Eureka! The superb results changed the course of Irvine-Halliday's career, and improved life for tens of millions of people.

> ". . . Dave Irvine-Halliday realized that a single 0.1-watt, white-light-emitting diode supplies enough light for a child to read by. The simple but revolutionary technology supplied to homes by his **Light Up The World Foundation (LUTW)** can light an entire rural village with less energy than that used by a single, conventional, 100-watt lightbulb."
>
> —*2002 Rolex Award for Enterprise*

LEDs' efficiency provides energy-constrained, low-income villagers and slum-dwellers the best bang for their lighting buck:

- ✔ No energy is lost through heat, advantageous in hot climates.
- ✔ LEDs use just one-sixth of the energy of an incandescent bulb and two to three times less than CFLs (compact fluorescent lights).
- ✔ LEDs typically are guaranteed for 25,000 hours, or twenty years of usage.
- ✔ They can be dimmed to conserve available energy.
- ✔ Because LEDs provide so much light with so little energy, the solar chargers to power them can be small, adding to their affordability.
- ✔ They can be miniaturized, allowing for small task-lighting bulbs consuming just tiny amounts of energy.

LED bulbs, like cell phones, leapfrog over earlier iterations straight to cutting-edge technology. Other selling points:

- ✔ Bright, aesthetically appealing white light
- ✔ Instant illumination, with none of the start-up lags common to many CFLs

Lighting Africa's mission is to implement industrywide LED standards. Low-quality lights that fail are termed "**market spoilers**"; they sow potential customer distrust. Lighting Asia is also performing LED testing and setting performance standards. Standards motivate manufacturers to produce higher-quality products.

YOU

- Check **Light Up The World (LUTW)** for current volunteer openings. "Volunteers are a valued resource at LUTW. Over the years we have had countless dedicated individuals who have invested their time and passion in technical support, IT, training and research, administration, community awareness, and fundraising."

Solar Home Systems

Rooftop solar panels provide residential power to turn on the lights, charge mobile phones, and power small TVs.

GreenEmpowerment.org • @grnempowerment

Carrying solar panels to Humla, a remote Nepal village.
© Jodi Winger, courtesy of Bradley Hiller

Low-income countries' power grids will not extend to most remote areas for decades—if ever—consigning many of the world's rural poor to dark, kerosene-lit nights and a constant scramble to power mobile devices. But a distributed solution is available.

One roof-mounted, 75-watt photovoltaic (PV) solar panel allows a low-income rural family to leapfrog over central grids. When fully charged, the panel powers four lights (or more, if they are LEDs) and a TV, and charges cell phones, while costing less for energy than before. Solar home systems are becoming more economical:

- ✓ Solar panel prices are dropping.
- ✓ Kerosene, battery, and diesel prices are rising.
- ✓ LED bulbs (**#25**) use very little electricity to provide ample lighting.

Solar home systems extend families' productive time into the evening. Non-smoky, reliable reading light enables children to read and do homework. Women can more effectively perform both domestic and income-producing work, and socialize more with neighbors. For artisans, more light means more production time. Home-based cell phone charging saves both time and money.

Bangladesh's **Grameen Shakti** ("Village Energy"), a pioneer in solar energy, sold its first Solar Home System in 1995. In 2012, it turned on the power for customer number 1 million. Growing from a single branch office to more than 1,000, Grameen Shakti continually refines its products based on feedback gathered through its famously attentive customer service. Payment-plan options are twenty-four or thirty-six months. Households can reallocate the money formerly spent on kerosene and batteries toward paying off their solar loan.

Shakti also supports a microgrid program for its lowest-income customers. Owners of solar systems sell their surplus energy to neighbors who cannot afford the setup.

Solar home systems:

- ✓ Provide clean, high-quality light
- ✓ Eliminate kerosene's smoke-induced respiratory and eye irritation and disease
- ✓ Lessen fire and burning hazards
- ✓ Become a long-term asset when paid off, generating nearly free electricity for up to thirty years
- ✓ Produce power passively—no daily remembering to put it outside. It is up there on the roof, charging during daylight hours
- ✓ Create local and global eco-benefits

Challenges:

- Systems require maintenance; owners need training in the basics.
- Energy output is reduced on rainy, cloudy days.
- On-site customer support is time-consuming, especially in sparsely populated regions.
- In cyclone-prone regions, panels should be removed before storms.
- In flood-prone regions, batteries need to be moved up high—in Bangladesh, to the rafters!

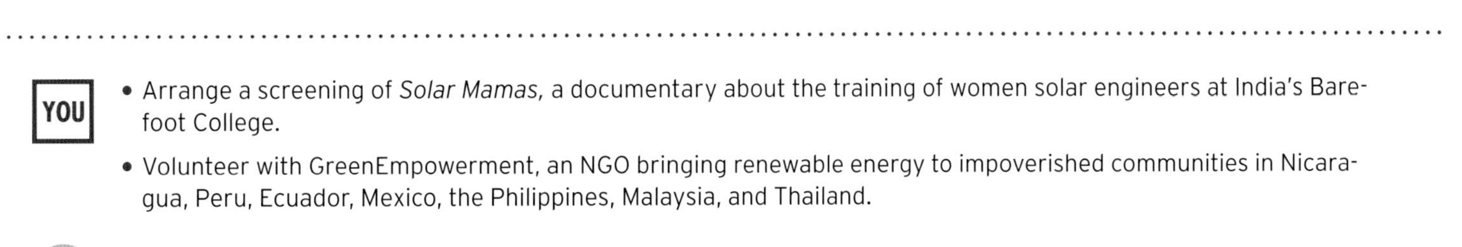

YOU

- Arrange a screening of *Solar Mamas*, a documentary about the training of women solar engineers at India's Barefoot College.
- Volunteer with GreenEmpowerment, an NGO bringing renewable energy to impoverished communities in Nicaragua, Peru, Ecuador, Mexico, the Philippines, Malaysia, and Thailand.

EDU

- Read Nancy Wimmer's *Green Energy for a Billion Poor*, chronicling the development of Grameen Shakti.

School girls work on their arithmetic problems, in Peru. © GVEP/EMPRENDA

Portable Solar Lanterns and Lamps

Portable solar LED lamps and lanterns are increasingly affordable replacements for kerosene, providing superior light while eliminating health hazards and fuel costs.

Luminet.org • @SolarSister.org • @Solar_Sister

27 Finally, low-cost portable solar lamps are coming on the market, entry-level lights with the potential to improve quality of life while saving money. Kerosene's feeble flames are the usual light source for the estimated 1.3 billion people without access to grid electricity.

A fossil fuel, kerosene emits smoke and toxic fumes, harming the people who are directly exposed, as well as the Earth's atmosphere. Kerosene has many additional negatives:

- ✗ A constant risk of fires and burns, especially in densely populated neighborhoods with houses constructed of highly flammable materials.
- ✗ Danger of explosion.
- ✗ Smoke inhalation causes respiratory and eye irritation and life-threatening disease.
- ✗ Time and effort is expended to travel to kerosene purchase points.
- ✗ Lamps are frequently extinguished by wind.
- ✗ Kerosene light is dim, barely enough to read by.
- ✗ Kerosene fumes are unpleasant.

Many families in low-resource areas spend as much as a fifth of their income on candles or kerosene. That's a lot to spend on inferior, smelly light. Solar-powered lamps are dropping in price. If financing is available, solar lights can quickly be paid off with the savings from eliminating kerosene and flashlight-battery purchases.

High-quality evening lighting provides many dividends:

- ✔ Schoolchildren can read and do their homework.
- ✔ Women can use their evenings more productively.
- ✔ Business owners can expand shop hours.
- ✔ Shopkeepers avoid needing to clean soot from their merchandise caused by kerosene smoke.
- ✔ Lamps enhance household security.
- ✔ Portable lights make women's night outings less dangerous.
- ✔ Solar lanterns are inflation-proof—sun is free while the cost of fuel is rising.
- ✔ Midwives have created ingenious head-mounted solar lamps, illuminating nighttime deliveries and improving maternal outcomes.

> "An elderly woman who rolls *beedi* (tobacco leaves) for a living spoke about how the constant fear of the tobacco leaves catching fire from the kerosene lamp slowed her down. Earlier she used to roll 5,000 beedis a week and now she's able to increase that to 7,000, and as a result earn Rs250 [$4] more every week."
>
> —*Milaap, an Indian person-to-person microfinance organization*

Dozens of companies are producing lamps and lanterns in the $10–$20 range, consisting of:

- ✔ Freestanding solar-charging panels
- ✔ LED bulbs
- ✔ Standing or hanging units housing the storage batteries

A typical daily charge yields eight to ten hours of light. High/low switches allow users to conserve energy supply. In response to the enormous penetration of mobile phones, higher-end lamps now frequently include cell phone charging.

The potential demand is enormous, not just in rural settings but also in peri-urban slums. Solar lights are more reliable than those powered by unreliable electrical grids with daily outages. Grid-connected consumers purchase them for backup lighting, too. New models can even charge on cloudy days.

Manufacturers continually upgrade product lines, lowering prices and improving performance. **D.Light** has sold more than a million units and estimates customers have saved nearly half a billion dollars in energy expenses.

Challenges:

- In a rush to bring inexpensive solar lamps to market, some are manufactured without replaceable batteries. When batteries die, the lanterns become electronic trash.
- Inferior products are market spoilers, sowing distrust among potential customers. Many manufacturers now include a one-year warranty. Initiatives like Lighting Africa raise product standards.
- If owners forget to put lamps out to charge during sunlight hours, there is no light that night.
- Solar lights and lanterns improve health and productivity at affordable prices, but customers need access and familiarity. Paid off solar lamps give free light for several years, except for the cost of replacing a battery.

SOLAR SISTER

Thirteen-year-old Amina Ibrahim Abdullah of the UNHCR Camp Doro Mabaan, South Sudan, holds her Luci solar LED lantern by Mpowerd. © Sebastian Rich

Uganda-based **Solar Sister** is an innovative social business providing both training and start-up loans for its solar entrepreneurs. Demonstrating and promoting solar products, their sales force reaches out to their own communities. Who knows better than local, connected women about all the benefits offered by these newfangled lights?

Recognized by their colorful branded uniforms, Solar Sisters sell many different companies' products. Solar Sisters develop skills in maintaining and repairing their product lines, building trusted relationships with their customer base. Since they are local, their expertise stays local, too. They support themselves and their families, while extending the benefits of clean, inexpensive solar lighting to their communities.

Solar lanterns are leapfrogging technology, bringing users of 19th-century kerosene lamps straight to 21st-century, clean, solar light. Skipping dirty, fossil-fueled, CO2-spewing power plants and leaky transmission lines altogether, customers will never need to purchase inefficient incandescent bulbs or unpopular CFLs. The benefits will be both local and global.

See Also: **LEDs** (**#25**), **Liter of Light** (**#76**), **LuminAid** (**#34**)

According to their analysis, every dollar invested in a Solar Sister Entrepreneur generates more than $46 in economic benefits in the first year alone, through earned income for the Solar Sister Entrepreneur and the immediate cash savings of her customers, as they avoid the cost of expensive kerosene.

YOU

- **www.ElephantEnergy.org** works in Navajo Territory in the American Southwest, where many residents have no grid electricity, as well as in Namibia. They feature a **Buy One Give One** solar light program.
- Donors can finance a **Solar Sister's Business in a Bag** through their revolving loan fund. Once the loan is paid off, the money is lent to bring on a new team member.

Eden Full, the SunSaluter's inventor. © Della Robbins

SunSaluter

The SunSaluter facilitates solar panels' rotation from east to west, generating 40 percent more solarization than if the panel were simply lying flat.

D·I·Y H L&G tel

SunSaluter.com • @TheSunSaluter

Large solar farms program their solar collectors to rotate from east to west, which maximizes sunlight exposure over the course of the day; domestic solar panels just sit there. Peak solarization only occurs at noon when the sun is indeed overhead. British Columbian Eden Full noticed this problem as a young teenager. An inventor by temperament, she quickly started searching for possible fixes.

The result is Full's innovative **SunSaluter**, an add-on to a conventional solar panel. With time out for a Thiel Fellowship, Full is now a student at Princeton University majoring in mechanical engineering.

SunSaluters, constructed from simple, widely available materials, rotate solar panels from east to west, increasing their power charge by 40 percent without consuming any electricity. Accomplished by water displacement, a low-tech process, the panel rotates in an east-west arc over the course of the day.

- The solar panel balances on a wooden trestle with a hinged central beam for support.
- Two inverted bottles of water are suspended from the east-facing end of the panel. Tubing is attached and the flow rate adjusted so the bottles' contents empty over the course of the day, collecting in a clean water receptacle.
- The bottles have filters to purify the water, an added benefit. Those who lack grid connection generally also lack treated water. (If users trust SODIS (**#39**), filtration could be eliminated.)
- A counterweight is suspended from the west edge of the panel. As the eastward water empties, the western weight gradually lowers the panel until, at day's end, the panel slants toward the west.

In an early pilot in a Kenyan village, a woman complained that their charger powered just two of their three solar lamps. Full promised to return with hardware to help build a SunSaluter. When she looked for the woman a few days later, she was nowhere to be found. To Full's dismay, she learned the woman had ventured out at night without a light, searching for firewood, and been trampled to death by a water buffalo, leaving behind two children. For Full, the task of helping people overcome energy deprivation became much more personal and urgent.

Challenges:

- Daily labor is required to set the system up in the morning. Maximizing its benefit means getting up with the sun.
- Nightly removal of the clean water is required.

SunSaluter has run pilots in Tanzania, Mexico, and Uganda. Eden Full is a popular speaker, keynoting events highlighting innovation and humanitarian tech for poverty alleviation.

EDU

- Build your own SunSaluter. SunSaluter is eager to spread the word about the increased solar efficiency of their innovation and is available to consult on educational projects.

Pay-Till-You-Own Electricity

Solar panels for off-grid, nonelectrified households can be installed and paid off in weekly installments via mobile phone.

tel

Quetsol.com • @_quetsol • Ashden.org • @AshdenAwards

Electric lights come on for the very first time when twelve-year-old Amanda punches in the code, Peru.

Pay-Till-You-Own is a contemporary installment plan. Integrating mobile payments with electricity generated on site, by solar panels or wind turbines, permits families to access the benefits of clean energy with only a down payment for the system. Weekly fees cover a week's quantity of energy, as well as a fraction of the system's total cost. After a fixed number of payments, systems are unlocked and customers own them.

Simple metering, **Pay-As-You-Go**, also utilizes mobile money. While benefiting from the clean electricity and saving on fuel expenses, customers don't purchase the equipment.

Electronic payments lower overhead costs for suppliers, but companies must provide customer credit. Pay-Till-You-Own integrates:

- On-site renewable energy generation
- Mobile payments (**#90**)
- Microcredit provided by the energy company or a partner (**#88**)

Users' costs for batteries, kerosene, and cell phone charging are radically lowered or eliminated, with resources reallocated toward weekly prepayments. When paid off, the system continues to generate nearly cost-free electricity for many years. Customers also:

- ✔ Access the health benefits of clean energy
- ✔ Create a digital credit history, helpful for future transactions
- ✔ Save the time formerly spent procuring fuel and charging phones

Quetsol is turning on the lights in rural Guatemala. A certified **B Corporation**, Quetsol integrates **triple bottom line** values: people, profits, and planet. Over 2 million Guatemalans lack grid connection; Quetsol seeks to redress this energy inequity.

Azuri, a Kenyan pay-to-own solar company, reached an important milestone in 2013 when its first round of customers completed payment. Anne became a happy owner, noting: "Now that I have unlocked, I am the envy of my neighbors —they cannot wait until it is their turn." Azuri received Ashden's Award for Innovation, presented by the British NGO, to encourage greater use of sustainable energy in reducing poverty and tackling climate change. Azuri's **Energy Escalator** offers bigger systems powering more lights, radio, TV, and even sewing machines.

Challenge:

- Since each microgrid is independent, customers only can access as much energy as their microsystem produces (true for all distributed energy).

Takamoto Biogas, based in Nairobi, uses cell phones for electronic payments for gas produced by portable, above-ground biodigesters installed on small farms. Customers fill the units with animal waste. Methane gas is produced and used for cooking and other energy needs.

Lumeter offers a platform for cell phone payment metering for any product. The system photographed above is a wind-turbine, activated by the Lumeter keypad. Lumeter technology also can be used for solar panels, TVs, or even electrical water pump services, turning any electronic appliance into a pay-as-you-go device.

- **Sunfunder.com** is a for-profit investment platform for pay-to-own solar systems in Uganda.
- **Quetsol** offers sponsorship opportunities for individuals, foundations, and businesses.

Shelling maize using Global Cycle Solution's bike-powered sheller, Arusha, Tanzania. © Philemon Kivuyo

Bike-Powered Machines

Bicycles can leverage human power to run a wide variety of specially engineered machines, performing tasks quickly without consuming electricity or fuel.

mayapedal.org • @mayapedal • GCStz.com • @gcsTZ

In energy-poor settings, bikes (**#78**) don't just provide transportation. Human pedal power can run machines for farms, homes, and small businesses. Two approaches are:

- Guatemalan **Maya Pedal**'s reengineering of donated bikes into pedal-powered machines, **bicimáquinas** ("bike machines").
- Designing tools that attach to functioning bikes, **Global Cycle Solution**'s strategy.

Bicimáquinas cost well over $100, but are typically used by cooperatives that share costs and benefits. Partner NGOs collect and ship donated bikes to Maya Pedal's San Andrés Itzapa workshop. The workshop supports itself by retooling and selling these bikes along with their signature bike machines.

Bicimáquinas are suited for small farms and businesses for use with:

- Water pumps
- Grinders
- Threshers
- Roof tile makers
- Nut shellers
- Blenders (for making soaps and shampoos, as well as food products)

"We are in the process of making our designs available globally, via downloadable fact sheets and step-by-step instructions. We aim to be a center of pedal power research and development and an information resource for NGOs promoting appropriate technology and small scale, sustainable agriculture."

—*Maya Pedal*

Global Cycle Solutions, headed up by MIT-trained mechanical engineer Jodie Wu, is based in Tanzania. One of their most popular offerings is a bicycle-powered cell phone charger. The device charges a phone when attached to a bike during a ride, or when users simply pedal the bike while it's stationary.

Global Cycle Solutions is prototyping a **Universal Bicycle Adapter**. "The universal bike adapter is a high-impact tool which, upon installation to any bicycle, instantly gives the user affordable access to a suite of human-powered agricultural technology. The adapter harnesses bicycle power without altering the structure or functionality of the bicycle as a vehicle," Wu explains.

The company also markets a $60 detachable maize sheller that does the job ten times faster than individuals, paying for itself in a month. Corn-shelling can become an income-generating initiative when users ride the bike to where shelling is needed.

©TED

"Because the foundation of our business was unique in leveraging the bicycle to create simple, affordable, income-generating technologies for the bottom of the pyramid consumer, I knew I had to start Global Cycle Solutions, no matter the challenges ahead."

–Jodie Wu, CEO

YOU

- Maya Pedal works with volunteers organizing the collection and shipping of used bicycles. They also need writers, photographers, and translators to create instruction manuals, as well as long-term volunteers.
- **Global Cycle Solutions** is a seeking **Impact Investors** to provide capital for expansion. GCS also has fellowship opportunities in Tanzania.

Biogas Backpack

Biogas backpacks provide a low-tech method for biogas, produced in a central biodigester, to be transported for use as cooking fuel.

 tel

Be-nrg.com

Katrin Puetz's invention for biogas transport, Guder, Ethiopia.
© Katrin Puetz

As a young development worker in Africa, German-born Katrin Puetz became convinced that social enterprise was the model to pursue to bring people modern, clean energy. She focused on biogas production using biodigesters.

Biodigesters are an elegant, well-established technology. Usually built underground, they transform waste into methane gas and slurry, a semi-liquefied fertilizer, providing a sanitation solution for treating waste and superior fuel piped directly to nearby stoves.

> "Cooking with gas" now describes tasks that are quickly accomplished. Originally, it was an advertising slogan touting the virtues of gas stoves over their wood-burning predecessors.

Biodigesters require at least a small farm's quantity of waste to be productive. They are popular in many other regions around the world, each one a locally constructed biogas mini-plant. Puetz has developed an Ethiopian biogas business model, with farmers delivering waste to a centralized, more technically advanced biodigester. Farmers pick up a gas allotment when they drop off; surplus gas is sold.

Puetz dreamed up the low-tech biogas backpack to overcome the distribution hurdle of transporting small quantities of gas. Despite its expansive volume, 1.2 cubic meter of biogas —a full backpack—weighs less than ten pounds. When filled with gas, they are safe, but as a precaution they should remain outdoors, attached by a flexible hose to an indoor stove.

Puetz sought culturally appropriate, locally available materials to construct the backpacks. She spotted a plastic-lined fabric honey transport bag in the Addis Ababa markets, a promising lead.

Back in Germany, Puetz discovered airtight, heavy-duty dunnage bags, for filling empty cargo ships' open spaces, which she repurposed for biogas transport. At present the backpacks' cost is in the $35 range.

Puetz's company, **(B)Energy**, provides end users with many biogas benefits:

- ✔ Waste management solution
- ✔ Decreased flies and insects attracted to waste
- ✔ Elimination of the time and effort expended foraging for and chopping wood
- ✔ Lowered deforestation
- ✔ Clean-burning fuel, with almost no smoke, radically decreasing smoke-induced eye and respiratory disease and irritation
- ✔ Fertilizer to improve soil, lowering commercial fertilizer expenditure
- ✔ Slurry can be used as fish food (**#73**)
- ✔ Local jobs marketing and delivering biogas

Katrin Puetz, the biogas backpack inventor, receiving the 2013 Siemens-Stiftung "Empowering People" Award.
©Georgina Goodwin/Siemens Stiftung

(B)Energy will provide revolving microloans for biogas entrepreneurs. Biodigesters generally pay for themselves within two to three years, and last for decades.

See also: **Microcredit** (**#88**)

YOU

- This biogas business model is scalable and replicable. Entrepreneurs looking for opportunities should contact **(B) Energy** to help build a network by launching elsewhere.
- **(B)Energy** welcomes inquiries from **Impact Investors**.

Catarina Lopez of Fundación Progresar with EcoLeña eco-briquettes
Sacapulas, Guatemala. © Peter Stanley/LegacyFound.org

Eco-Briquettes

Eco-briquettes — fuel fabricated from waste products — burn cleaner, create local jobs, and preserve forests.

D·I·Y H L&G tel

LegacyFound.org

Eco-briquettes might look like mud pies, but they are serious business, creating high-quality, clean fuel from waste. At the bottom of the energy ladder people meet their fuel needs by burning dung, crop residue, and wood. While readily accessible, these are extremely inefficient energy sources and emit toxic smoke, making open-fire cooking dangerous and unhealthy.

Briquetting mixes:

- Combustible waste (in some cases filling a waste management gap)
- Water
- Binder (usually clay, but starch or gum resin from local trees also works)

Combustible ingredients can include:

- Sawdust
- Bagasse—sugarcane or sorghum waste
- Waste paper
- Coffee husks
- Straw
- Charcoal dust particles too small to sell

The excess water is extracted and the mixture is compacted, shaped, and air-dried. More surface area enhances burning performance, hence the holes in the middle.

Eco-briquette composition is flexible. Briquettes can be made by hand. Simple machines, ratchet presses, require a modest investment but create higher-quality briquettes, increasing quantity as well as profits.

Using eco-briquettes with high-efficiency improved cookstoves (**#50**) offers multiple benefits:

- ✓ Freeing up women's and girls' wood-gathering time for more constructive pursuits
- ✓ Relieving women's physical strains from hauling heavy loads of wood
- ✓ Alleviating pressure on forests and providing global carbon capture
- ✓ Utilizing waste materials
- ✓ Offering local microbusiness opportunities, often run by women

As cooking fuel, eco-briquettes outperform firewood because they:

- ✓ Ignite more quickly
- ✓ Generate more heat
- ✓ Produce less smoke, decreasing eye and respiratory disease and irritation
- ✓ Deposit less soot on pots, saving domestic scouring labor

As a child growing up in Kenya, Dr. Mary Njenga was inspired by Nobel Laureate Wangari Maathi's environmental stewardship. By focusing her doctoral research on briquette tech, analyzing ratios and contents to create the highest-efficiency briquettes, Dr. Njenga was able to lower fuel expenses while helping the environment.

Dr. Mary Njenga studies the combustion qualities of briquettes. © Robin Chacha

- ✓ In Kibera, a slum of Nairobi that Dr. Njenga studied, householders who produced their own eco-briquettes saved a whopping 70 percent on fuel. Those who purchased ready-made briquettes still saved 30 percent compared to conventional fuel purchases.
- ✓ In Nepal, with assistance from the **Foundation for Sustainable Technologies (FOST)**, the Jharuwarashi orphanage in Katmandu, Nepal, makes its own eco-briquettes. Their firewood expenses have decreased 30 percent, as has their carbon footprint.

Legacy Foundation, based in Ashland, Oregon, works with groups in more than thirty countries, providing information and sharing knowledge on eco-briquettes.

YOU

- Legacy Foundation welcomes engagement in its work through volunteer activities, internships, and idea-sharing.

Green Charcoal: Biochar Briquettes

An Eco-Fuel Africa entrepreneur, Uganda. © EcoFuelAfrica

Biochar-based briquette production utilizes crop waste to produce high-quality, inexpensive cooking fuel and fertilizer while decreasing deforestation.

 tel

EcoFuelAfrica.org.ug • @EcoFuelAfrica1 • CharcoalProject.org

Sanga Moses' ticket out of rural Uganda was education; he became an accountant at a large Kampala bank. On a visit home, he was distressed to see his kid sister coming around the bend, loaded down with firewood. It was pointless for him to pay for her schooling, she tearfully told him, because she missed too many days collecting firewood to keep up with her lessons.

Moses had collected firewood in his youth a decade earlier, but trees had been more plentiful and closer to the village then. Massive deforestation had turned fuel-gathering into a full-day task. He devoted himself to finding a sustainable solution.

Moses zeroed in on **biochar-based briquettes**. Conventional charcoal is made by carbonizing wood in kilns. Biochar also uses **carbonization**, but utilizes agriculture waste, sparing trees. Farmers formerly burned crop residue to get rid of it. For biochar briquetting, this waste is carbonized in repurposed oil drums.

Biochar briquettes offer end users:

- ✔ Money pocketed, costing 20 percent less than carbonized wood charcoal
- ✔ Smoke-free burning, making cooking healthier and more pleasant than wood or charcoal
- ✔ Long-lasting flame, designed for local stoves (**#50**)
- ✔ Time saved scrubbing pots, because it leaves less soot residue
- ✔ Time saved by eliminating wood gathering duties, helping children stay in school

Eco-Fuel Africa, the social business Sanga Moses built, features an integrated supply chain from field to kiosk.

- ✔ Biochar is sourced from a network of local farmers, providing them with income, as well as biochar for use to improve their soil's quality and productivity (**#57**).
- ✔ Employees use Eco-Fuel Africa's manual hydraulic **Eco-Fuel Presses**, designed for off-grid settings.
- ✔ Briquettes are delivered by teenagers on bicycles (**#78**), providing yet more employment.
- ✔ Branded kiosks are staffed by marginalized village women—typically divorcees, single mothers, and widows—recruited by Eco-Fuel Africa to sell Biochar Briquettes and whatever additional sundries they choose.
- ✔ Promoting habitat restoration, Eco-Fuel Africa has planted more than 150,000 trees (**#69**) by distributing seedlings to schools and community groups.

> "We have selected women who have at least one daughter, and agree with them that part of the income from the sales of our products should go toward financing of the daughter'(s) education."
>
> —*Sanga Moses*

Eco-Fuel Africa has expanded by setting up microfranchises (**#92**). The business has excellent potential: Fuel is a daily necessity, and there is an abundance of agricultural waste in Uganda.

YOU

- Check with Eco-Fuel Africa about volunteer and internship opportunities.
- Eco-Fuel Africa, a social enterprise, seeks impact investments.

EDU

- Watch **Amy Smith**'s famous TED Talk, Simple Designs to Save A Life, viewed nearly 700,000 times. Smith explains biochar briquetting, a project of MIT's D-Lab that she founded and heads. Instructions are at their website, **www.d-lab.mit.edu**.

Sector 4 – An Overview

WASH: Water, Sanitation, and Hygiene

For those who turn faucets and—voilá!—potable water pours out, on-demand water access is unremarkable. Such people don't need to:

- Find the water
- Transport the water
- Disinfect the water
- Store the water

For a billion other people in the world who lack access to clean water, these are all issues of pressing daily concern.

Similarly, for those with rich infrastructure:

- Hand washing and toileting are routine, not daily challenges that require advance planning.
- Diapers are stocked, dirtied, and laundered—or tossed, if disposable.
- Garbage cans are rolled to the street, their contents hauled away to remote landfills.
- Recycling systems are in place.

About 2.5 billion people on earth lack toilets, latrines, and waste management systems. Their excreta pollute shared water sources, spreading diarrheal disease in a cycle perpetuating ill-health and killing children.

Women and girls worldwide are tasked with providing water and maintaining household hygiene. The scene of biblical Rachel at the well is reenacted daily by millions upon millions of women and girls who spend hours each day drawing and carrying water. At least Rachel had a well; many in the contemporary world don't. Journeying to open water sources and carrying home jerry cans—forty-four pounds of water—is wearying for them, and even harder when they're also lugging babies.

It is not possible or desirable to replicate 19th-century sanitation systems. It consumes immense resources to use clean water to flush waste, build sewer systems to move it to costly treatment plants, add inputs to decontaminate it, and then discharge what is left into waterways. A much smarter, more eco-friendly approach is to harvest energy and fertilizer from waste through **eco-sanitation.**

Likewise, it is too costly to treat water centrally and deliver it. Many affordable home-based water treatments accomplish this task efficiently and effectively.

WASH investments yield enormous dividends. Development experts agree water and sanitation hygiene is enormously impactful, improving health and productivity and paying back each dollar invested many times over. Although it is culturally and technologically complex to improve water access, sanitation, and hygiene, low-hanging fruit exists, and is described in this section.

The absence of WASH services is bad for everyone but harms women in specific ways:

- ✗ Women's time and energy is spent on water provision.
- ✗ Women are responsible for water treatment. This is time-consuming, and if they treat water by boiling over open fires, smoke exposure is itself a health hazard.
- ✗ Women, lacking privacy, often use the bushes at night, making them vulnerable to sexual assault.
- ✗ Absence of private toilets and waste disposal systems result in many girls being absent from school during menstruation and often dropping out at menarche.
- ✗ Women bear the burden of caring for children sickened by water-borne diseases.

Affordable WASH solutions have the potential to radically improve women's productivity and quality of life, raising up their families and national economies with them. The rising tide of improved water and sanitation hygiene **will** lift all boats.

34. DISASTER RELIEF: LIGHTS AND CLEAN WATER

Disaster relief has two new compact and lightweight tools: inflatable lights and water-transport backpacks used with water-purification tablets.

35. RAINWATER HARVESTING

Collecting and storing rainwater provides water security and safe drinking water, sparing women from hauling their family's water supply. Surplus water can be used for irrigation.

36. IMPROVED WATER TRANSPORT

Wello's WaterWheel™ doubles water transporting capacity with decreased effort. Pack H_20 is an ergodynamic upgrade for carrying water.

37. POINT-OF-COLLECTION WATER CHLORINATION

Randomized controlled studies indicate that point-of-collection chlorine treatment facilitates broader compliance than home-based, end-user chlorination.

38. WATER-TESTING KITS

Inexpensive water-quality testing allows locals to monitor and manage water sources and provide their community with up-to-date water quality info.

39. SOLAR DISINFECTION

Placing untreated water in plastic bottles and leaving them in the sun for six hours kills the pathogens, producing safe drinking water.

40. SODIS INDICATORS

Gauges and indicators visually indicate when solar water disinfection (SODIS) has been accomplished, building end-user confidence.

41. SOLVATTEN SODIS/HOT WATER HEATER

The Solvatten features easier water transport and speedier solar water disinfection, with a built-in safe water readiness indicator.

42. WATER FILTRATION DEVICES

Water filtration disinfects water by straining out pathogens through porous ceramic or layers of sand, trapping microbes.

43. SOLAR WATER DESALINATION

The Eliodomestico solar-powered desalinator produces drinkable water from salty seawater.

44. HANDWASHING

Humble handwashing packs a serious punch, lowering disease transmission, slowing the spread of outbreaks, and saving lives.

45. HUMAN WASTE ASSET MANAGEMENT

PeePoople has developed waste collection bags in which excreta are transformed into fertilizer.

46. LATRINES: ECO-SANITATION

State-of-the-art eco-sanitation provides safe waste collection while utilizing waste's nutrients, transforming them into methane gas and fertilizer.

47. MENSTRUAL SUPPLIES

There is a growing consensus that providing menstrual pads improves girls' school attendance. Locally manufactured pads help meet growing demand.

48. DIAPERING

The absence of affordable diaper options means infant and toddler poop often falls under the sanitation radar, contaminating local water sources.

49. GARBAGE RECLAMATION

In regions with no municipal garbage services or recycling, residents often provide informal waste reclamation services.

A wondrous inflatable LED light arrives in Haiti. © LuminAID

Disaster Relief: Lights and Clean Water

Disaster relief has two new compact and lightweight tools: inflatable lights and water-transport backpacks used with water-purification tablets.

CSDW.org • @PG_CSDW
luminAID.com • @LuminAIDlab

34 Two innovative products intended for disaster use, both designed by women, provide creative solutions for providing light and clean water in emergency situations. LuminAID lights are paired in this entry with water treatment, though they could also be included with LED solar lights (**#27**).

Disaster management requires speed and appropriate supplies to provide shelter and infrastructure for survivors of natural catastrophes, such as earthquakes, floods, and typhoons, as well as refugees of war and conflict.

The disaster management sector, a coalition of governments, intergovernmental organizations under the UN umbrella, and independent NGOs like the Red Cross, has vast expertise. Disaster supplies are funded by donors. Shelter, meal provision, medical care, sanitation, water, and communications must be up and running within hours or, at most, days. End users often arrive with just the clothes on their back, unable to pay for services.

LuminAID co-founders Anna Stork and Andrea Sreshta, architecture students, shared an interest in solar humanitarian tech. After Haiti's massive 2010 earthquake, they launched LuminAID Lab.

LuminAID lights pack flat and weigh just eight ounces, lowering both the cost of transport and the space required. Fifty inflatable LuminAID lights fit in the space taken by a box of eight flashlights.

The lights are translucent, diffusing the harsh beams of the LEDs inside. Though designed for disaster use, they would be popular anywhere people need modern, well-designed, off-grid lights.

Tricia Compas-Markman designed a disaster relief water purification solution as a student at California Polytechnic State University, where she founded its Engineers Without Borders chapter. Her innovative **DayOne Response Waterbag™** backpacks ship flat, saving weight and space. They were recently deployed to the Philippines, supplying Typhoon Haiyan victims with clean water.

The ten-liter bags are filled with water, carried to their intended locations, and suspended by built-in hooks. They have attached hoses for easy dispensing and are much easier to carry than the jugs and jerry cans they replace. Proctor & Gamble **PŪR®** water purification packets decontaminate the water in thirty minutes.

PŪR® water purification tablets, though effective and affordable, were a commercial flop. PŪR is frequently distributed in crisis settings. Each 3.5-cent packet utilizes coagulation and disinfection to transform ten liters of visibly grungy liquid into clear, potable water—similar to techniques used in large water treatment facilities. Strict adherence to PŪR treatment protocol is difficult in disaster relief settings, however.

See also: **Solar Lights and Lanterns** (**#27**)

YOU

- **LuminAID** supplies disaster relief NGOs like ShelterBox, and also donates lights to partner initiatives. LuminAID features "Give Light, Get Light" retail sales. For each retail purchase, a lamp is donated to one of their field partners.

EDU

- Test **PŪR®** out on dirty, debris-filled water and watch the magic.

Rainwater Harvesting

Collecting and storing rainwater provides water security, sparing women from hauling their family's water supply. Surplus water can be used for irrigation.

 tel

ProximityDesigns.org • @ProximityDesign • Rainsaucers.com • @rainsaucers

RainSaucer water collecting in Quetzaltenango, Guatemala. © RainSaucer™

Rain storage technology is ancient. Rainwater harvesting is a widely practiced water conservation technique in industrialized countries, augmenting municipal supplies. In regions lacking water infrastructure, stored rainwater is a major source of water for households and small farmers, especially during the dry season.

Women and girls are traditionally responsible for the family's water supply, and perform arduous daily water-sourcing treks, filling heavy receptacles and carrying them home. This becomes especially onerous during the hot, arid months. Wells and creeks run dry, forcing families to purchase water at great expense or pushing women farther afield to find water.

Rainwater harvesting and collection provides multiple benefits:

- ✔ Water security during dry seasons
- ✔ Freeing up women's time
- ✔ Minimizing or eliminating girls' and women's physical exertion and strain of water carrying – full plastic jerry cans weigh forty-four pounds
- ✔ Reallocating funds, if water was formerly purchased
- ✔ Water for irrigation, especially during the dry season

Rain storage can be as low tech as collecting rainfall by the liter, aided by funnels fashioned from bottles. Where rainfall is abundant, flexible water tanks (also called water bags, baskets, or bladders) can store sufficient water for dry season domestic usage and surplus water for irrigating. *Out Of Poverty* author Paul Polak is an enthusiastic rainwater harvesting advocate, having facilitated Asian subsistence farmers' drip-irrigation (**#64**) using collected rainwater. Adding a dry-season crop expands their income significantly.

Flexible water storage tanks, far cheaper than stationary cisterns, can be moved to take better advantage of water sources. Myanmar-based **Proximity Design** introduced a 250-gallon, $23 water basket in 2012. Made of PVC-impregnated canvas, it stores flat, growing into a freestanding tank as it fills with water. Add a pump and hose, and irrigation becomes a quick one-woman job.

RainSaucers™ are freestanding rain barrels fitted with large, flattened funnels to maximize water collection. RainSaucer is piloting an entry-level $2 collector made of plastic-covered corrugated cardboard. In Latin America, many citizens rely on bottled water. Collecting rainwater minimizes bottled water purchases, though stored water should be treated before drinking.

Rooftop rainwater harvesting is practiced the world over. Immense concrete rain jars are standard in rural Thailand. Costing around $25, they store nearly 850 gallons. Owners follow basic precautions to avoid contamination and collect free, natural drinking water. Bottled water costs 75 times more.

Corrugated tin roofs are excellent rain harvesters, with their ridges channeling rainfall. Inexpensive gutters can be fashioned from pipes or horizontally cut bamboo.

YOU

- **RainSaucers™** seeks NGO partners for its Two Dollar Tank rainwater harvester.
- **Proximity Designs** offers competitive internships for Myanmar-speaking high school graduates and college students. Fellowships are awarded to exceptional graduate students.

- Research and construct a water harvesting system for your school garden.

A WaterWheel™ user rolls her water load in Ahmedabad, Gujurat, India.
© Wello/Cynthia Koenig

Improved Water Transport

Wello's WaterWheel™ doubles capacity with decreased effort. Pack H2O is an ergodynamic upgrade for water carrying.

WelloWater.org • @wellowater
PackH2O.com • @PackH2O

One in seven people in the world live more than a kilometer from their closest water source. In villages, this means carrying empty containers to a well, or for those less fortunate, an open water source. Slum women line up at taps, hoping for an adequate supply. Readers of Katherine Boo's *Beyond the Beautiful Forevers* will recall that in Annawadi, the Mumbai slum of her lyrical book, the taps frequently run dry; waiting is in vain.

Piped water is not happening for the bottom billion in the foreseeable future. The path to improving the lives of hundreds of millions of girls and women is the design of affordable water transport devices. **Social entrepreneur** Cynthia Koenig is doing just that, having taken the **WaterWheel™** through multiple iterations, adhering to the principles of Humanitarian Tech Design:

- ✔ **Co-created** with end users
- ✔ Culturally acceptable
- ✔ Attractive
- ✔ Affordable, with the price dropping when manufactured in large quantities

Wello's WaterWheels™, with their fifty-liter capacity, hold more than double the water women carry in *matkas*, water containers common in India. Rolling easily on the ground, they free up at least one hand. (Hello, cell phone!) Fetching water in half the time yields an extra hour or two a day to be reallocated to more productive pursuits. Girls have more time for schoolwork, and don't arrive at class exhausted from having carried their family water for miles. Adult women can engage in income-producing labor or communal activities. WaterWheel™ user Varsha reports, "After hours of water collection every day, I used to be very irritable with the kids. Now I am a better mother." Wello customers procure an average of ten extra liters of water a day for drinking and improved hygiene.

> "While the WaterWheel™ was created with women in mind, as they tend to collect water, [inventor] Koenig says Wello has been surprised by its popularity among men. 'One of the most exciting things is that men love using it, they see it as a tool.'"
> —*The Guardian PovertyMatters Blog*

Wello's WaterWheel™ has won numerous design competitions, funding initial prototyping. Microcredit could provide financing to facilitate the purchase, projected at an affordable $25. It will also facilitate water-dependent microenterprises like:

- ✔ Cold drink vending
- ✔ Water delivery
- ✔ Tea stands
- ✔ Laundry services

Upgrades under consideration are a filter to initiate water disinfection during transit and harvesting energy from the wheel's rotation for cell phone charging services.

PackH$_2$O water packs lessen women's physical strain when carrying water. Designed to ergodynamically distribute the burden, they are an upgrade over jerry cans and buckets. PackH20s weigh one-seventh of a jerry can's weight, an improvement all by itself.

YOU

- **Wello WaterWheels™** is looking for partner NGOs and social businesses to expand their market.
- **Wello** welcomes volunteers.

Point-of-Collection Water Chlorination

Randomized controlled studies indicate that point-of-collection chlorine treatment facilitates broader compliance than home-based, end-user chlorination.

 tel

PovertyActionLab.org • @JPAL_Global • evidenceaction.org • EvidenceAction • ZimbaWater.com • @ZimbaWater

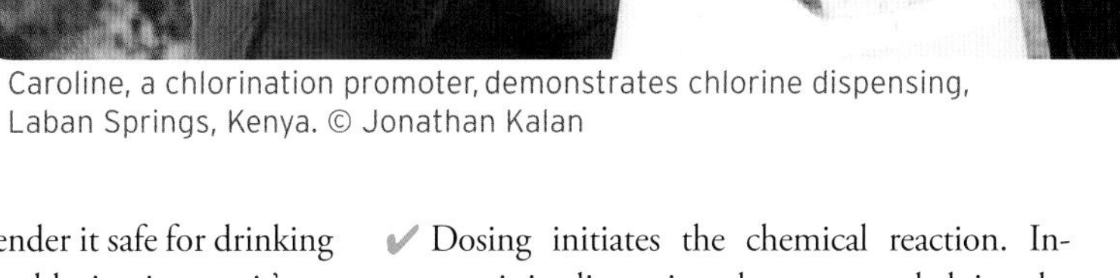

Caroline, a chlorination promoter, demonstrates chlorine dispensing, Laban Springs, Kenya. © Jonathan Kalan

Water treatment is a hard sell for villagers unfamiliar with germ theory. The idea of invisible pathogens lurking in their water causing their frequent illnesses is difficult to prove. With the added hassle and expense of water treatment procedures, motivating compliance is challenging. The industrialized world has solved this problem—pre-treated water is delivered directly to people's homes.

The **Poverty Action Lab (J-PAL)** studies poverty alleviation interventions to determine what actually works, helping to assure funding is used maximally effectively.

J-PAL seeks to identify effective water treatment compliance strategies. Clean water provides enormous health benefits at a very low cost. Waterborne illnesses drain energy and require medications. Sick children need extra care, diverting maternal time away from other tasks. What is the most effective way to motivate women to adopt recommended water treatment procedures?

Their studies show point-of-collection chlorination works far better than expecting people to treat their water once they've hauled it home from the central water source. **Chlorine Dispenser Systems** are an inexpensive poverty buster. Boiling water to render it safe for drinking is much more work than chlorinating, so it's an upgrade for women.

> Chlorine, Cl_2, atomic number 17, causes bacteria-killing chemical reactions when added to water. It is used to treat municipal drinking water and in chlorine bleach, bottled disinfectant. Chlorinated swimming pools' recognizable smell is due to the higher concentration required to maintain a bacteria-free expanse of still, exposed water.

A village woman is hired to maintain the Chlorine Dispenser, manage chlorine supply, collect households' fees, and promote healthy water habits. Added advantages of this system include:

- ✔ Bulk chlorine is significantly cheaper than individual packets.
- ✔ The valve releases an accurate, metered dose, more reliable than home-based mixing.
- ✔ A desire to conform to social norms motivates people to dose their water publically.
- ✔ Dosing initiates the chemical reaction. In-transit jostling mixes the contents, helping the chlorine do its job more quickly.
- ✔ Women are already collecting water, so dosing is just a small modification, easier than creating a new at-home habit.

Central chlorine-dispensing systems costs are just 50 cents per person, per year.

Zimba, an innovative, automatic chlorine doser, is in development at MIT's D-Lab, a center of humanitarian tech innovation. Now based in West Bengal, India, the Zimba is in final testing. Zimbas are powered by gravity and the force it exerts on water, delivering a chlorine dose:

- ✔ Without hinges or gears subject to malfunction
- ✔ Without electricity
- ✔ That's automatically calibrated based on the quantity of sensed water
- ✔ Costing around $6 per year per household when shared among families

Users don't chlorinate their water—Zimba does it for them.

YOU • **ZIMBA** is seeking distribution partners.

A student tests water quality in school lab, Mangalore, India. © BASF

Water-Testing Kits

Inexpensive water quality testing allows locals to monitor and manage water sources, providing their communities with real-time water quality info.

WomensEarthAlliance.org • @WomensEarthAlly • mWater.co • @mWaterCo

Until recently, water quality testing required expensive lab equipment and highly trained technicians. New testing technology has made water quality monitoring simpler, portable, and affordable for low resource communities where polluted water causes widespread illnesses.

Improving sanitation helps prevent water contamination. Contracting and retransmitting waterborne illnesses, the source of an estimated 40 percent of hospitalizations in the developing world and a major killer of kids under five, is a vicious cycle. Individuals infected with diarrhea excrete infectious pathogens. Without sanitary toilets or latrines, these are often reintroduced into community water sources, spreading yet more infection.

Water tests visually indicate the presence of infectious germs. They are easy for community members to interpret, making them an effective tool for teaching germ theory and motivational for mobilizing communities to improve their shared sanitation.

A number of affordable kits are on the market. Testers add the correct amount of water, wait, and check for a color change to know if water is contaminated or safe to drink. **Portable Microbiology Laboratory** kits, for example, contain:

- Whirl-Pak® stand-up bags for collecting water samples
- 10ml test tubes containing a substance that reacts with *E. coli,* changing color if it's present
- Petrifilm™ test plates that yield a count of the sample's bacteria colonies
- Sterile, disposable pipettes to transfer water from Whirl-Paks to the test tube and Petrifilm™
- UV flashlight to help detect *E. coli*

The German charitable foundation BASF Stiftung partnered with UN-HABITAT in launching a chemistry curriculum in Mangalore, India, a city with frequently unsafe water. Studying water protection, sanitation, and conservation, 5,000 students learned how to test local water quality and interpret results. The students were then deputized to put their knowledge to work, testing water in their home community.

Women's Earth Alliance trains women's groups in a wide range of WASH skills, including water testing. If tests show contamination, water monitors communicate with the local population, implement solutions, and work to create alternative safe water sources, if needed.

The **open-source**, free **mWater** app uses a mobile phone's camera to detect *E. coli* bacteria from water samples, grown on test plates. Data can be added to the larger knowledge base to view local, regional, or wider area water status. Users can subscribe to receive alerts and recommendations for alternative clean water sources.

> "In Tanzania, more individuals have access to mobile phones than access to safe water sources. Phones therefore provide a way of creating cheaper infrastructure for monitoring water."
> –*Anne Feighery, CEO* **mWater**

EDU

- Test water in your local watershed. In areas with advanced sanitation, the contaminants are generally chemical pollutants from industrial or agricultural runoff. A list of water-testing kits is posted at the **mWater** website.

Solar Disinfection

Placing untreated water in plastic bottles and leaving them in the sun for six hours kills the pathogens, producing safe drinking water.

tel

Sodia.ch • FundacionSodis.org • EAWAG.ch • @EawagResearch • WaterSchool.com • @WaterSchool

Reaching for water treated by solar disinfection, Tegucigalpa, Honduras. © Matthias Saladin/ Fundación SODIS

39 Waterborne diseases spread by untreated water cause millions of annual deaths and illnesses; it is shocking that free water treatment via sunlight, **SODIS**—an acronym of "solar disinfection"—is not more widely known and utilized.

SODIS has impeccable credentials; laboratory research in the 1990s confirmed its effectiveness. WHO, the UN, and CDC all recommend it. After six hours in the sun (sometimes even less; two days if it's cloudy), ultraviolet rays kill all viruses and bacteria, rendering the bottles' contents safe to drink.

Heating water to just 149°F/65°C kills all disease-causing bacteria, viruses, and parasites. Boiling to 212°F/100°C provides visual evidence of water safety, but the energy consumed in raising the temperature the extra 63° is consumed unnecessarily. The challenge: knowing when the water reaches 149°F/65°C.

What obstacles are preventing broader SODIS adoption?

- Lack of awareness is a major hurdle, though NGOs work hard promoting SODIS. The Swiss Federal Institute of Aquatic Science and Technology, **EAWAG**, takes a lead role, providing extensive educational resources on SODIS.
- Lack of user confidence slows SODIS uptake: Can something so simple actually work? SODIS gauges and indicators (**#40**) address this. Visual signs that the water has reached a high enough temperature help users trust this treatment technique.
- For larger quantities of water, filling and emptying bottles is unwieldy, since the optimal SODIS bottle size is two liters. EAWAG is introducing a four-liter hanging SODIS bag.

SODIS is a zero carbon-emitting treatment that consumes no fuel. For families disinfecting water by boiling it over wood-burning stoves, adopting SODIS yields significant savings of time, money, and indoor air pollution.

Studies show that when the SODIS method is applied correctly in #1 PET plastic bottles, there is no danger to human health from chemicals leaching into the water. People without access to treated water are less concerned about chemicals in plastic than with pathogens in their water, but it is reassuring to know the method is safe.

It takes two pounds of wood to bring one liter of water to a boil. Imagine how much timber is needlessly consumed for boiling water, generally gathered by girls and women. Eliminating wood-gathering duty frees up women and girls' time, and also curbs deforestation.

YOU

- Promoting an unbranded, free technique is challenging. Ironically, SODIS might catch on faster if a social enterprise designed, branded, and marketed attractive SODIS bottle sets. Check out **SolarBottle.org**'s design ideas.
- **FundacionSODIS.org** seeks volunteers with experience in management, finance, fundraising, research, communications, and information systems to work on Latin American SODIS projects.

- Test out SODIS. **Waterschool.com** has a comprehensive FAQ list.

Training in the use of WAPIs, Water Pasteurization Indicators, Jinja, Uganda. © Diane Parrish

SODIS Indicators

Gauges and indicators visually indicate when solar water disinfection (SODIS) has been accomplished, building end-user confidence.

IntegratedSolarCooking.com • @Potavida.org • @Potavida • @Helioz.org • @HeliozGmbH

SODIS (#39), solar disinfection of water, is accomplished by placing untreated water into clean plastic bottles and setting them in the sun for six hours (or two days, if cloudy).

Treating water using SODIS offers many benefits:

- ✔ Safe water saves lives by destroying infectious waterborne microbes, including *E. coli,* rotaviruses, giardia, and hepatitis A virus.
- ✔ Solar disinfection uses no fuel, preserving forests.
- ✔ Girls' and women's time, formerly spent gathering wood to boil water, is freed up for other pursuits.

WAPIs, Water Pasteurization Indicators, help convince people that SODIS works. Soy wax melts at 149°F, the same temperature required for treating water. WAPIs are small, clear, sealed plastic cylinders encasing a lump of green soy wax. When inserted inside the bottle, melted wax indicates completed water treatment.

Potavida, a SODIS gauge designed by a University of Washington team that provides "Safe Water with Certainty," won the competition posted by **Fundación SODIS**.

Potavida co-creation in Nicaragua. © Anna Young

The Potavida has since been through three iterations, its inventors working with end users to **co-create** a product incorporating their input and meeting their needs. Earlier designs (as seen in the photograph above) were for a bottle attachment, but Team Potavida's latest version integrates the UV gauge into a SODIS bag, having concluded this is a more intuitive and user-friendly approach.

The **WADI**™—short for "Water Disinfection"—was designed by **Helioz**, an Austrian social enterprise focused on humanitarian technology. It attaches to the top of a bottle placed horizontally for SODIS treatment. The WADI detects and measures the sun's UV rays. A solar-powered smiley face lights up when the water is safe for drinking.

In optimal circumstances, disinfection can occur much faster than the recommended six hours, or two days if cloudy. An advantage for gauge users is only leaving the bottles in the sun as long as is necessary. The amount of time required to accomplish disinfection depends on several variables:

- Weather conditions (clear, cloudy, rain)
- Intensity of the sun's rays
- Elevation (the higher, the faster)
- Latitude (the closer to the equator, the faster the process)

Helioz field tests in various regions and weather conditions around Asia and Africa have shown pasteurization takes from forty-five minutes to several hours. (Without a gauge, the full six hours should be allotted.)

- WAPI kits are available for 42 cents apiece when purchased in quantity, for assembling. Rotary Club of Fresno, California runs WAPI projects and collects finished WAPIs for distribution.

Solvatten SODIS/ Hot Water Heater

The Solvatten features easier water transport and speedier solar water disinfection, with a built-in readiness gauge.

tel

Solvatten.se • @Solvatten

Mother and daughter disinfect water with their Solvatten in Bungoma, Kenya. © Solvatten/Johanna Felix

Solvatten sounds like something you'd buy at IKEA, and that's not far off. Swedish for "sun water," it's a water treatment device invented by Swedish microbiologist and artist Petra Wadström.

Wadström lived in Indonesia and observed how much time and labor women and children spent transporting and treating water. She was sure there was a way to accomplish these tasks more effectively. Back in Sweden, she began an eleven-year process of designing the Solvatten. Elegant on many levels, it speeds the SODIS (**#39**) process while doubling as a water heater. Her invention:

- ✔ Is a hinged, ten-liter black box that opens flat, maximizing the surface exposed to sunlight
- ✔ Includes a built-in filter, easily removed for cleaning or replaced by several layers of an old sari (the filter catches debris, not microbes)
- ✔ Features a push-button indicator that turns green when internal temperatures have reached pasteurization level
- ✔ Weighs six pounds empty, around thirty-one pounds when filled with water
- ✔ Lasts five years or more, costing just $.002 cents per water liter

In direct sunlight, Solvattens purify water in two to six hours, making multiple pasteurization cycles a day possible. Users are warned that the water is very hot, over 149°; domestic water heaters are generally set between 120° and 140°.

Additional benefits include:

- ✔ Hot water to speed cooking and brew tea with no fuel consumed
- ✔ Improved school attendance, since clean water access improves girls' health and decreases their wood-gathering duties
- ✔ Hot water for personal hygiene

Solvatten has unleashed entrepreneurial creativity:

- ✔ A restaurateur in Kawempe, a slum of Kampala, Uganda, serves a free glass of Solvatten Water with each meal and explains why she no longer sells plastic water bottles. Her daily volume has increased from 50 meals to 70–100 meals in a few months.
- ✔ Two Ugandan women package Solvatten-purified water in Ziploc bags that they sell to local kiosks. Using two Solvatten units, on sunny days they can fill 180 bags, or 120 if it's cloudier, and earn up to $5 a day.

Wadström now heads Solvatten, an NGO working with partners in Haiti, Kenya, Nepal, and Uganda. Units are being distributed at around $35 through NGO partners.

Petra Wadström holds her award-winning Solvatten. © Catharina Hansson

Solvatten's role in preserving forests has qualified it for **Gold Standard Carbon Offset Credits**.

See also: **SODIS** (**#39**), **SODIS Indicators** (**#40**)

YOU

- Start an American Friends of Solvatten to make it easier for Americans to support their work.

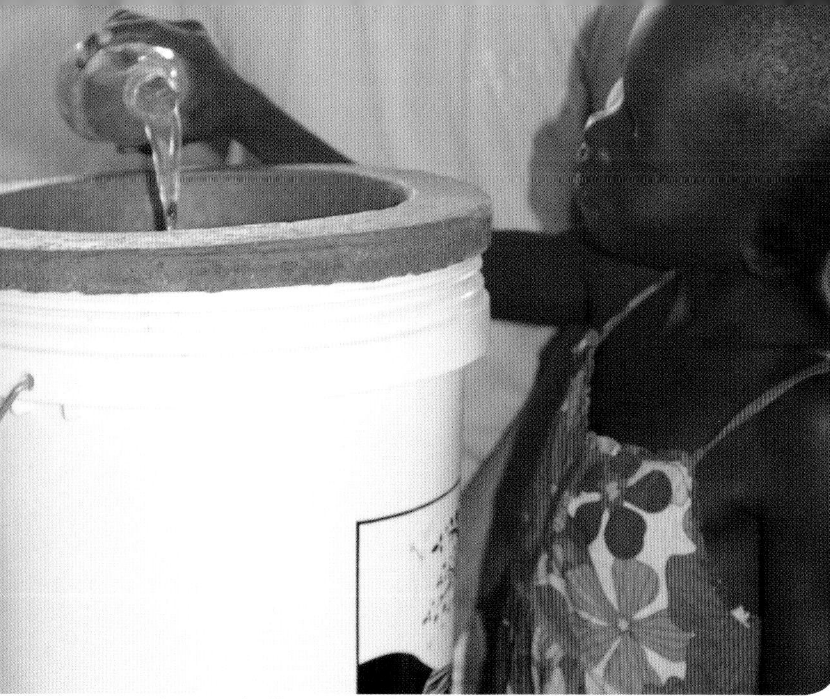

Learning how to filter water with Spouts of Water, started by Harvard students in 2012, Kumi, Uganda. © SpoutsOfWater.org/Suvai Gunasekaran

Water Filtration Devices

Water filtration disinfects water by straining out pathogens through porous ceramic or layers of sand, trapping microbes.

H L&G S2N tel

PottersForPeace.org • @PottersForPeace

Ceramic filters were designed in Guatemala in 1981 by Dr. Fernando Mazariegos. Most communities already create and use clay vessels, making ceramic filtration a culturally appropriate WASH technology.

The flowerpot-shaped ceramic filtration unit sits inside a larger, lidded plastic urn with a spigot. Untreated water is added from the top; it filters through the ceramic and collects in the urn. Treated water is protected from new contamination, flowing from the urn when the spigot is lifted.

Factories are producing ceramic filters in more than twenty countries, based on procedures developed and disseminated by **Potters For Peace**. Potters for Peace chooses not to patent their internationally recognized design. They post it online, encouraging **open-source** communication and innovation. Ceramic filtration devices cost around $10.

✓ Pots are manufactured from local clay, keeping material and transportation costs low.
✓ Artisans are local, creating jobs.
✓ Factories employing four potters can produce fifty ceramic filters a day.
✓ Clay is mixed with sawdust or other particulate that burns off during the firing process, leaving myriad microscopic traps for pathogens.
✓ Pots are molded to a uniform size holding 8–10 liters of water.
✓ Fired pots are coated with colloidal silver, a microbicide.

Water filtration often replaces boiling water, which consumes wood or charcoal fuel. In 2012, Ashden presented its Green Energy Award to IDE-Hydrologic, a Cambodian ceramic water filter manufacturer. Honored for their filters' role in forest conservation and climate change mitigation, they have qualified for **carbon offset funding**.

Challenges:

- Filtration is less effective at killing viruses than it is at killing bacteria.
- Filtration takes several hours.
- Equipment needs to be cleaned periodically to avoid recontamination.
- Ceramic is, of course, breakable. Cracked filters need replacing.

Biosand Water Filtration was developed in the 1990s by a Canadian, Dr. David Manz. Biosand filtration devices are vertical concrete or plastic bins. Inside them, strategic layers of gravel and sand gradually filter out untreated water's pathogens. Potable water comes out a tube. They are constructed with local materials by local labor, providing jobs along with improving health. Units cost about $50, and last up to thirty years. The design is patented but available free online.

In 2013, eighteen-year-old Megan Shea's experiment utilizing moringa seeds (**#69**) to screen pathogens from contaminated water snagged her a spot as a finalist in the Intel Science Talent Search. Moringa trees are plentiful in the developing world.

YOU

- Join the **Potters for Peace Brigade** on their annual two-week winter working trip to rural Nicaragua, working with local potters and spending a day at a ceramic water filter factory.
- Potters For Peace seeks volunteers with strong grant-writing or computer skills to help expand its reach.

Solar Water Desalination

A solar-powered desalination device produces fresh water from salty seawater.

 tel

GabrieleDiamanti.com • @GabDiamanti

Harvesting clean, potable water from sea and sun. © Gabriele Diamanti

Humans have dreamed of desalinating seawater forever. With population growth expected to put increasing pressure on water supplies, desalination accomplished via free sunlight could provide a vast new freshwater source. Current desalination tech consumes large quantities of fossil fuel, solving one problem but exacerbating another, and costs too much to be poverty-alleviating.

97.5 percent of the earth's water is salty ocean water. Were that water easily, affordably, and sustainably desalinated, it could help meet people's drinking and sanitation needs and also expand irrigation.

Cities tend to cluster along coastlines; large urban populations could satisfy at least some of their water needs through desalination. It would be particularly useful for island communities contending with finite water supplies.

Freelance Italian designer Gabriele Diamanti has won a number of prestigious awards for his **Eliodomestico**, a solar distiller producing 5 liters of clean water daily from salty or brackish water.

The United Nations recommends that for cooking and drinking alone, people need a minimum of about 5 liters; lactating mothers require 7.5 liters. The Eliodomestico is a good start, especially if it supplements other sources of safe water.

Diamanti's distiller works like an upside-down coffee percolator, capturing steam and directing it downward into a collection pan.

1. In the morning, seawater is poured into the black ceramic boiler evaporator sitting atop the still. The top is screwed tight to prevent the steam from escaping.
2. Solar heat is absorbed until pressure builds.
3. When a high enough temperature is reached, the water boils and is converted to steam, filtering out salt.
4. Condensed steam is captured and piped into the bottom collector pan, producing up to 5 liters of water.
5. The water is also disinfected, having reached temperatures higher than necessary (149°) to kill pathogens.
6. Fresh water is removed and transported home.
7. Since all the salt has been filtered out, iodized salt may be added to the distilled water for health and taste purposes. If the water source was not contaminated with chemicals, the salt can be harvested.
8. If the water is exposed to pathogens in transit, it may need additional treatment.

As of this book's writing, Diamanti is completing the Eliodomestico's third design iteration. He will post **open-source** design plans online, available for anyone, anywhere wanting to produce them. His attractive, low-maintenance design utilizes:

- ✔ Locally available, inexpensive materials
- ✔ Techniques familiar to developing world potters
- ✔ No chemical or filter inputs
- ✔ No moving parts
- ✔ No electricity or batteries

Diamanti will encourage local production, lowering costs and providing livelihoods. The price is estimated at $50, requiring subsidies or credit for the extremely poor. He will manufacture and distribute the product from his studio in Italy, too, to promote the Eliodomestico and raise funds to disseminate it.

A lesson in handwashing with an inverted bottle Tippy Tap; the screw cap is loosely attached. Touching it releases a few drops of water, Cochabamba, Bolivia. © Avery White/Proyecto Horizonte, Bolivia

Handwashing

Humble handwashing packs a serious punch, lowering disease transmission, slowing the spread of outbreaks, and saving lives.

TippyTap.org • @TippyTap_org • GlobalHandwashing.org • @HandwashingSoap

Handwashing? Surprising as it may be that a tool this simple helps achieve poverty alleviation, handwashing with soap is one of the most powerful public health interventions around. But it is challenging in low-resource settings lacking water taps.

Tippy Taps, low-cost, low-tech, water-conserving handwashing devices, were designed in the 1980s in Zimbabwe by Dr. Jim Watt. Consisting of a suspended water container with nearby soap, most have a manual foot pedal to tip the container, releasing water hands-free. Tippy Tap variations have proliferated.

Educational campaigns stress handwashing to limit the spread of bacteria, viruses, and general germs:

- After using the latrine
- Before preparing food
- Before feeding a child
- After cleaning a baby

The Glitter Ball Game is a fun technique for teaching about germs. A ball smeared with petroleum jelly is dipped in glitter. Tossed to and fro, the glitter sticks all over participants' hands. To remove it, they must use soap and scrub hard. Such it is with germs.

Personal hygiene FAQs:

Is hot water necessary?

No. Studies show room-temperature water works just as well, with the added advantages of not drying out skin or wasting energy to heat water.

Is soap necessary?

Yes. "Using soap to wash hands is more effective than using water alone because the surfactants in soap lift soil and microbes from skin and people tend to scrub hands more thoroughly when using soap, which further removes germs." – CDC

Is shampoo necessary?

No. A growing movement in the industrialized world favors the "no poo" approach, eschewing shampoo as chemicals that strip hair of natural oil. Some use just water or baking soda to gently clean hair. In the developing world, shampoo is sold by the packet to low-income customers for whom it is a luxury item.

How do people shower without running water?

A vast majority of the world's poorest live in very hot climates with no air conditioning; hot showers are not necessarily desirable. Common methods are bucket showers or bathing in rivers (suboptimal, since river water is likely polluted.)

Bio-Sanitation Centers (**#46**) are a promising approach for delivering sanitation services to those without plumbing. Bio-San centers feature handwashing, clean latrines, and private showers on a fee-per-use basis. Waste is collected in a central biodigester (**#33**) that generates methane gas, used to heat the water.

Happy Global Handwashing Day–October 15

YOU

- **Proyecto Horizonte** welcomes Spanish-speaking volunteers to assist in their integrated health/education/community development programs promoting SODIS (**#39**) and handwashing.

EDU

- Build a Tippy Tap Exhibit based on designs used around the world.

Human Waste Asset Management

PeePoople has developed waste collection bags in which excreta are transformed into fertilizer.

 tel

PeePoople.com • @PeePoople

PeePoo bags in Kibera, a Nairobi slum with no municipal sanitation. © PeePoople/Camilla Wirseen

Poop smells, and it harbors viruses, bacteria, worms, and parasites that are easily spread via water or hands. But human waste has value: its high nutrient content. The industrialized world uses large quantities of potable water to flush excreta from toilets to sewers. Vast amounts of resources are then consumed treating the sewage with chemical inputs to render it safe for discharging into open waterways. No value is captured.

Rural and slum populations in low-resource areas are searching for methods for safe, affordable human waste disposal accomplished without constructing large infrastructures. Containing waste safeguards water, averting the endless cycle of water being polluted by human waste and then re-spreading illnesses in the community. **Eco-sanitation** system designs:

- ✔ Contain and decontaminate human waste using a minimum of water and/or chemical inputs
- ✔ Recapture waste's assets, yielding valuable by-products

Biodigesting (**#31**) accomplishes this by collecting waste in closed chambers where it decomposes anaerobically, producing methane gas for cooking as well as slurry, liquefied fertilizer. However, the waste volume required is larger than individual households produce. It is an effective communal solution, the centerpiece of bio-san latrine centers (**#46**).

In many communities, "night soil," poop collected in buckets, is used as fertilizer. Without first allowing for safe decomposition, this is a dangerous, disease-spreading practice.

Rural populations without access to latrines meet nature's call by going in the bushes, referred to as open defecation. In mega-slums, hundreds of people often share poorly maintained latrines. Sewage-filled streams run straight through crowded communities. Bagged feces—"flying toilets"—are tossed in ditches. Obviously there is an acute need for better approaches.

> "The average economic benefit of a $1 investment in sanitation is $9.10."
> *—WHO, 2007*

PeePoos, invented by Anders Wilhelmson, a Swedish architect and urban planner, are affordable, biodegradable, single-use plastic bags that fit over standard buckets. Users are charged 3 cents per bag and paid 1 cent when they return them full to central collection points. Fees are lower than public latrines charge. Full bags turn into high-nitrogen fertilizer within a few months. PeePoo bags are micro-treatment plants:

- ✔ No water is consumed.
- ✔ Bags contain urea, which turns into ammonia when it comes into contact with urine and feces. Ammonia destroys disease-causing bacteria within a few days.

In addition to keeping waste out of the local waterways, thereby improving public health, PeePoos:

- ✔ Provide at-home privacy
- ✔ Save time waiting in latrine lines
- ✔ Improve women's safety, avoiding night use of public latrines, a frequent venue of gender-based violence
- ✔ Decrease the presence of flies and other disease-spreading insects attracted by feces
- ✔ Provide jobs and income to vendors selling the bags

See Also: **Latrines** (**#46**), **Urine Fertilizer** (**#62**)

YOU

- **PeePoople** works with NGOs focused on slums, education, refugees, and emergency response. Inquire about partnerships.

Vellayammal Bodi with her first latrine, Ponnusangampatti, Tamil Nadu, India. © Water.org/Marla Smith-Nilson

Latrines: Eco-Sanitation

State-of-the-art eco-sanitation provides safe waste collection while utilizing waste's nutrients, transforming them into methane gas and fertilizer.

SuSanA.org • @susana_org • saner.gy • @Sanergy • oursoil.org • @SOILHait

One in three people in the world, around 2.5 billion, do not have consistent access to toilets or sanitation. Serious attention and effort is going into solving this immense problem, including the Bill & Melinda Gates Foundation's "Reinvent the Toilet Challenge" to create a toilet that:

- ✔ Sanitizes waste without sewer connections
- ✔ Uses no water
- ✔ Captures usable byproducts
- ✔ Costs users 5 cents a day, maximum

There are numerous approaches to affordable, user-friendly, ecologically and economically sustainable latrines in use.

The **Arborloo**, an entry-level latrine, is a 1.5- to 2-meter pit covered with a plastic latrine slab and sheltered by a lightweight structure. When the pit is nearly full, the slab and shelter are moved to a newly dug pit. Six to twelve inches of dirt are placed in the initial pit and a tree is planted, nourished by the composting waste.

Sustainable Organic Integrated Livelihoods (SOIL), based in Haiti, builds urine diverter toilets with fifteen-gallon drums under the seats. Full receptacles are removed through a side door and transported to a central composting site. Diverting urine lessens the weight of the collected waste, lowers transport costs, and reduces smell. Urine requires little or no treatment (**#62**); it filters into the surrounding area, fertilizing an adjacent garden.

Sasha Kramer, who earned a doctorate in ecology from Stanford University, is CEO and founder of Haiti-based SOIL. © Jon Brack

> "Sasha's work with SOIL is unique in that it is one of the largest and most promising tests of the paradigm-shifting hypothesis that sanitation no longer needs to focus on waste disposal, but rather on the economically profitable nutrient capture and agricultural reuse of human waste."
>
> —*Schwab 2014 Entrepreneur of the Year citation*

A for-profit latrine hospitality sector has emerged in densely populated slums. Cheerful, attractive, branded latrines like Kenya's **IkoToilet** are a major upgrade. People are willing to pay small fees to use them. **Eco-Sanitation Centers** feature a block of private latrine stalls, emptying into **biodigester** chambers (**#31**) that turn excreta into usable methane gas. Adjacent space is rented to businesses selling sundries and food, and services like mobile banking, shoe-shining, and barber shops, creating a community hub.

Social entrepreneur David Kuria, IkoToilet founder, emphasizes design, stressing that if latrines are beautiful, people will take good care of them. Indeed, IkoToilet Sanitation Hosts—coveted jobs—take pride in maintaining clean, branded IkoToilets.

Nairobi-based **Sanergy** features a branded **microfranchise** (**#92**) sanitation model. Individual latrines are managed by Fresh Life Operators who collect fees and maintain cleanliness. Fresh Life Toilets, locally pre-fabricated, are:

- ✔ Easy to keep clean
- ✔ Small enough to squeeze into densely populated spaces
- ✔ Affordable but include upgrades like hand washing and solar lighting
- ✔ Emptied daily, by wheelbarrow, since informal settlement roads are too narrow for vehicles

Aggregate waste is transformed offsite into renewable energy and organic fertilizer. Payment schemes range from single use to monthly family membership.

"Look what we built!" Schoolgirls from the Chisungu School with their new latrine in Epworth, Zimbabwe. © Peter Morgan/SuSanA

> **Composting toilets generally require the addition of biodegradable plant material or sawdust after each use, cover material that eliminates smell, reduces flies, and assists in the composting process.**

Equipping schools with proper latrines familiarizes children with safe sanitation practices, reinforces new habits, and specifically improves girls' school attendance. Some schools build latrines as a school community activity. **Tippy taps** (**#44**) located near the latrines make proper hand washing easy.

Latrine **microloans** (**#88**) are coming into use. Some microfinance institutions, expanding beyond their original limitation of lending money only for income-generating microenterprises, now offer home infrastructure upgrade loans, like **Solar Home Systems** (**#26**).

Cambodia has as active sanitation lending sector, whereby families seeking to invest in a latrine can borrow the money to finance construction. **VisionFund**, a Cambodian Microfinance Institution, serves as an intermediary, paying latrine construction enterprises directly. Purchasers pay their loans back to VisionFund while enjoying the benefits of home-based sanitation, essentially using an installment plan.

IDE Cambodia's "**Easy Latrines**," designed by Jeff Chapin, are popular:

- ✔ Made of local materials, they cost about $35.
- ✔ Sales reps are called Sanitation Teachers, promoting both hygiene and latrines.
- ✔ Slabs feature a choice of tile colors and patterns, attractive and easy to clean.
- ✔ When the receptacle fills up, a new receptacle is installed. Full drums are put aside for a year and become valuable compost.
- ✔ Latrine construction and installation businesses are booming. Many customers purchase latrines without subsidy or loans.

Additional economic and health benefits of hygienic latrines include:

- ✔ Home-based latrines decrease women's exposure to gender-based violence, especially after dark.
- ✔ In some locations, women who use the bushes at dawn or dusk are highly vulnerable to snake bites; latrines are much safer.
- ✔ Time is saved by less frequent illnesses, along with minimizing women's time spent tending to the sick.
- ✔ Toilet-hunting time is liberated. WHO estimates people spend about a half an hour a day searching for a private place to relieve themselves.

See also: **Biodigesters** (**#31**), **Urine Fertilizer** (**#62**), **Microcredit** (**#88**), **Microfranchising** (**#92**)

YOU

- Lend to a **Sanergy Fresh Life Operator** through **Kiva.org** to expand her sanitation microfranchise.
- **SOIL**, based in Haiti, welcomes help from skilled graphic designers, web programmers, and people with IT experience. Contact: volunteer@oursoil.org

A Rwandan farmer, Angelique Karidi, extracts banana fiber for SHE sanitary napkins. © Sustainable Health Enterprises

Menstrual Supplies

There is a growing consensus that menstrual pad provision improves girls' school attendance. Locally manufactured pads help meet growing demand.

DaysForGirls.org • @DaysForGirls • Afripads.com • @AFRIpads • SheInnovates.com • @SHEnterprises • Azadi.co.in • @AzadiCo

47 Many folk beliefs and customs about menstruation circumscribe girls' and women's freedom of movement and involvement in public life during their periods, unfairly isolating them and limiting their opportunities.

Women use mud, leaves, rags, or even newspapers to absorb flow. These makeshift solutions are ineffective, uncomfortable, and can cause infection. Girls avoid school for fear of staining and the resulting humiliation. Interventions that improve young adolescent girls' school attendance include:

✔ Reproductive health education, both for girls and boys
✔ Separate girls' toilets
✔ Discreet disposal systems
✔ Free menstrual supplies

Innovators are using the presentation of well-designed menstrual supplies as an attention-capturing moment to teach about puberty and menstrual management. If this takes place in primary school, girls are prepared for the onset of menses.

> "Teaching reproductive health is almost as important as the kits themselves."
>
> —*Days For Girls*

Menstrual pad options include disposables and reusables. In traditional cultures, vaginally inserted tampons and menstrual cups are considered inappropriate for schoolgirls.

Disposable pads, generally preferred over reusable cloth pads by schoolgirls, have many positives:

✔ Comfort and ease of use
✔ No washing or drying required, helpful where water access is limited and/or the climate is damp
✔ Protection from unwanted male attention (when pads are hung out to dry, they broadcast girls' maturity, putting them at risk)

The downsides of disposable pads:

- Monthly cash outlays required by governments or NGOs; schoolgirls cannot usually afford them.
- They are less eco-friendly than reusable pads

A number of social enterprises have sprung up, utilizing local materials, producing pads far cheaper than imported products, and creating local jobs throughout their supply chain.

In Uganda, Dr. Moses Kiiza Musaazi of Makerere University in Kampala has designed disposable **MakaPads** fabricated from local papyrus. Many MakaPad employees are refugee women. One of their customers is UNHCR, the UN's Refugee Agency, purchasing pads for refugee camp distribution. They also supply **OneGirl.org**, an Australian NGO, with pads for girls in Sierra Leone.

JaniPads, created by Swedish design students, are made from water hyacinth, an invasive, water-absorbing plant. That absorbency is an asset for menstrual products—a green, local solution. **Village Volunteers** has taken on the project and is seeking funding.

Sustainable Health Enterprises, **SHE**, manufactures disposable pads made from processed banana fiber, widely available agricultural waste. Based in Rwanda, SHE's new factory provides local employment both in manufacturing and marketing. Costing just 5 cents a pad, they are sold in bulk to donors who distribute them for free. They also sell pads retail, via microfranchisees.

Reusable, well-designed absorbent cloth pads also offer advantages, as they:

✔ Cost much less than disposables, over time
✔ Last many years
✔ Are eco-friendly
✔ Provide girls a reliable, independent method for handling a recurring event

Reusable pads' disadvantages include:

- They need laundering, which is time-consuming and requires soap and clean water.
- In damp climates or during rainy seasons, they can mildew before drying.
- Hanging them to dry where publicly visible can endanger girls.

Attractive cloth pads are both manufactured in the Global South and lovingly stitched by volunteers around the world.

Uganda-based **AfriPads** employs more than sixty workers, sewing cloth pads for distribution in schools.

Over 45 **Days For Girls** sewing clubs in six continents educate members about the challenges girls face in the developing world. Members then stitch up colorful pad sets for distribution through NGO partners. Their user-tested designs incorporate a barrier layer; kits include underpants and soap. Patterns are available at their site. Volunteer seamstresses can work independently of a club.

> "I tell people I am well past using sanitary pads anymore, but now my house is full of them."
>
> —*Gloria Buttsworth, Australia Days For Girls Club Leader*

It isn't just low-income school girls who need good-quality, affordable sanitary napkins. Modern women have more menstrual periods during their lifespans due to:

- Better nutrition, associated with earlier menarche
- Delayed marriage, extending the number of menstrual cycles before childbearing
- Dropping fertility rates; fewer pregnancies means more menstrual cycles
- Fewer pregnancies also means less breastfeeding; nursing suppresses ovulation/menstruation
- Lower breastfeeding rates (**#1**); mothers who bottle feed menstruate sooner

Combined with upward mobility, this creates greater demand for menstrual hygiene products.

Many Indian women's self-help groups are launching sanitary napkin manufacturing microenterprises using the machine invented by Arunachalam Muruganantham. His award-winning innovation produces 120 napkins per hour from local materials at one-third the cost of imported pads. More than 600 machines in 23 Indian states have been deployed, each one generating income and jobs, and improving women's health and quality of life.

Chopping banana fiber at the Ngoma, Rwanda SHE factory. © Sustainable Health Enterprises/Tash McCarroll

Azadi Pads is based in Uttar Pradesh where men own 85 percent of kiosks and are uncomfortable selling feminine products. Women's groups are selling them instead, filling the niche.

See also: **Diapering** (**#48**)

- Help **VillageVolunteers.org** promote their water hyacinth pad project: **Empowering Women Period**.
- Days For Girls club members become informed, committed advocates for global girls' and women's empowerment. Join or start a sewing club through **DaysForGirls.org**.
- **LunaPads.com**, an American cloth pad company that sells to the American market, donates one AfriPad to a Ugandan girl for each item purchased through their **Pads4Girls** initiative.

- Download **Menstrual Hygiene Matters**, a comprehensive resource book from **WaterAid.org**.
- Consult Dr. Linda Scott's Uganda menstrual management research at **DoubleXEconomy.com**.

Women's Global Toolkit
CHALLENGE:

Add a Sustainable, Affordable Diapering Solution HERE

Diapering

The absence of an affordable diapering option means infant and toddler poop often falls under the sanitation radar, contaminating water sources.

Scant attention is paid to how infant and toddler waste is managed in low-resource communities. The general assumption that kids go bottomless and toilet-train early with little effort, since the poorest billion mostly live in hot climates, is somewhat romanticized. Given the high level of infant mortality caused by sanitation deficits, surprisingly little effort has been made to promote diapering as a public health upgrade.

Toilet training is a misnomer when there are no toilets. Urine is not generally hazardous, but children's feces are pathogen-loaded. Children use potties, but this does not solve the problem unless the potties are emptied into safe latrines.

Latrines are unsafe and scary for small children due to the size of the opening. **Innovations for Poverty Action** has pioneered a children's latrine mat, **Safe Squat™**, placed on top of a latrine. Kids like them; mothers save time cleaning up after their children.

Chinese infants' britches have back slits facilitating frequent potty sitting. With one child per family, vigilant parents may manage early toilet training based on this conditioning. Elimination Communication adherents use non-verbal clues to anticipate babies' excretions, eschewing diapers. There is only a small body of anecdotal evidence, though, that this is what developing world mothers do, though bare-bottomed children generally toilet train earlier than kids wearing super-absorbent disposable diapers.

Mothers use makeshift cloth diapers, but without waterproof barriers, they leak.

In low-income countries, imported disposable diapers are luxuries sold one or two at a time. Sometimes mothers let them dry and reuse them.

Green Mama blogger Manda Aufochs Gillespie winters in Guatemala with her young children. She reports mothers there, immensely curious about her children's diapers, would love better diapering solutions.

She envisions a social enterprise producing a popular triple-decker diaper system:

1. **Diaper covers**, locally-sewn, with elasticized legs holding the inner diaper pad in place and avoiding leaks. Many designs and patterns are available online.
2. A **waterproof barrier layer**, integrated into the diaper cover. **PUL**, polyurethane laminated fabric, is widely used for this purpose.
3. **Inner diaper pads**, cloth or disposable, used interchangeably. Absorbent reusable cloth pads would be locally sewn. For disposable diaper pads, papyrus, water hyacinth, and banana fiber are inexpensive materials used successfully for manufacturing sanitary napkins (**#47**). They might make effective disposable diaper pads, too.

A diapering social enterprise will need to create a market by educating potential customers. While diapers are an aspirational status symbol, they aren't just an upscale consumer item; they prevent the spread of deadly pathogens when properly discarded or, ideally, composted. They also give mothers and other caretakers more flexibility and save a great deal of mess, liberating time that can be reallocated to other pursuits.

- Local diaper enterprises could provide employment, improve health and sanitation, and make life easier for the world's mothers — a trifecta of benefits. Start one!

Garbage Reclamation

In regions with no municipal garbage services or recycling, residents often provide informal waste reclamation services.

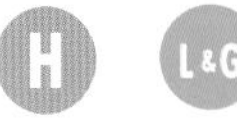

tel

Wiego.org • @WIEGOglobal • WeCyclers.com • @realwecyclers • Byoearth.com • @Byoearth

Waste pickers in Pune, India. © Julian Luckham for WIEGO and Waste Pickers Trade Union KKPKP

Rural areas and urban informal settlements rarely provide municipal garbage collection, leaving residents with few waste management options. Some cities dump all their garbage in landfills, creating immense, poorly managed sanitation hazards that breed disease and pests.

The world's poor have long been informal trash pickers, imperiling their health while performing valuable services. Waste pickers often achieve impressively high recycling rates, saving cities money in lowered landfill fees and conserving resources, an environmental benefit.

Women in Informal Employment: Globalizing and Organizing (WIEGO) represents legions of women toiling daily in the world's garbage dumps. One WIEGO initiative helps waste pickers achieve secure, improved working conditions. Bogotá, Colombia, recently recognized *recicladores,* waste recyclers, as official municipal employees, doubling their earnings.

An innovative incentive initiative, the **Garbage Premium Insurance Clinic Program**, conceived of by Gamal Albinsaid, has kicked off in Indonesia. Low-income enrollees pay 83-cent monthly microinsurance (**#91**) premiums—with garbage they collect and deliver, including organic waste—weighed on-site for composting.

Bilikiss Adebiyi-Abiola co-founded **WeCyclers**, a garbage recycling business, while at MIT's Sloan School of Business. Back in Nigeria, she heads this social enterprise in Lagos, a city of 18 million people. WeCyclers collects garbage with cargo bikes, well-suited for densely populated slums inaccessible to vehicles. As a contracted service provider to the Lagos State Waste Management Authority (LAWMA), they provide benefits to residents and the city both:

- ✔ Slum-dwellers receive points for recyclables, redeemable for cell phone minutes, food, and household goods.
- ✔ WeCyclers text public service messages to participants.
- ✔ Costs are covered by sorting aggregate materials and selling them to recycling processors.

A ByoEarth vermiculture demonstration, Guatemala. © ByoEarth

Maria Rodriguez, CEO of the award-winning Guatemalan social enterprise **ByoEarth**, approaches Guatemala's waste management deficit by enlisting worms to compost garbage, thereby decreasing the immense volume of landfill. Raised on a coffee plantation, Rodriguez was familiar with the need for eco-friendly waste management and organic compost to build up the soil.

ByoEarth takes their worm business in several different directions:

- ✔ Processing waste as a for-fee service
- ✔ Marketing worm-generated compost
- ✔ Selling domestic worm kits to cut back on household waste and generate usable compost
- ✔ Providing school demonstrations and classes on vermiculture and worm composting

Based in Guatemala, they deliver their fertilizer to plantations and farms, a closed-loop business generating income out of waste.

YOU

- **ByoEarth** welcomes interns. They recommend combining Spanish language school with part-time volunteering, helping ByoEarth grow their business.

Sector 5 – An Overview

Domestic Technology

The domestic behavior of affluent women around the world is endlessly scrutinized in order to design products for this coveted demographic to purchase. Not so for the world's poorest women. They have very little spending power. Scant attention has been paid to designing appropriate labor-saving technology for this demographic.

Given the strictly gendered roles of traditional culture, few men are attuned to designing technology for the domestic sphere. In many cases they simply know very little about women's work. Improved cookstoves (**#50**), however, are a major focus of design and marketing—hundreds of designs are in production, continually reiterated to improve their function—providing a trifecta of virtues:

1. Health benefits, from minimizing smoke exposure
2. Economic benefits, from decreasing fuel consumption
3. Ecological improvements, from conserving forests and decreasing **black carbon** emissions

Solar cooking (**#51**) scores even higher, with zero fuel burned and no smoke emitted. Thermal retention cooking (**#53**) saves even more fuel. Two additional benefits of solar and thermal cooking, rarely mentioned, are:

- ✔ These cooking technologies are cleaner, saving women a great deal time of cleaning time.
- ✔ Because food doesn't burn, pots don't need to be watched or stirred, freeing up women's time.

Given the energy poverty of women in low-resource areas, labor-saving tools can't rely on electricity. Several are featured here (**#54**), like nonstick pans, which save clean-up time, and corn shellers, which speed up manual labor.

Laundry is a largely ignored time-consuming endeavor. In the industrialized world, washboards are charming, antiquated musical instruments. In low-resource regions, women labor over dirty clothes the same way they have for centuries, in open waterways or by lugging water and bending over basins to use those washboards. The Giradora (**#56**) is a promising laundry innovation.

People value not only function but also beautiful design. The creators of LuckyIronFish (**#54**) learned that offering a block of iron for insertion into cooking pots had no appeal. When they reconfigured the iron into an attractive fish, a good luck symbol, acceptance was very high.

As electricity supply and incomes increase, there will be opportunities for marketing small appliances. Grinders, choppers, and mills will speed up tedious, repetitive tasks.

Domestic labor-saving devices liberate women. The time dividend can be invested in a variety of ways, including, but not limited to, income generation. More socializing, more cultural and educational opportunities, or more rest would all be appropriate uses for women's newly found time.

Let's pay more attention to how women go about their daily tasks and find ways to help them do them faster, with less taxing physical labor. Hopefully the ***100 Under $100* Second Edition** will have more domestic tech success stories to share.

50. IMPROVED COOKSTOVES

Clean Cookstoves consume less fuel, minimizing health-damaging smoke and black carbon emissions, and allow women to spend less time gathering wood.

51. SOLAR BOX STOVES

Wooden box solar stoves accommodate most cooking needs and provide economic, health, and environmental benefits.

52. PARABOLIC SOLAR STOVES

Parabolic solar reflectors concentrate the sun's rays, bringing the contents of a suspended kettle or supported pot to a boil more quickly than other solar cookers.

53. THERMAL RETENTION COOKING

Thermal retention cookers are insulated containers holding preboiled pots of food. The contents continue cooking without additional fuel.

54. IMPROVED KITCHEN EQUIPMENT

Better-designed equipment saves women time and drudgery, improving their productivity and quality of life.

55. COLD FOOD STORAGE

Evaporative coolers, made of ceramic and wet sand, keep produce fresh for many days, increasing farmers' and vendors' incomes.

56. LAUNDRY

Hand laundering is physically demanding and time-consuming. Scant attention has been paid to making this task less burdensome.

Jude Lally, an off-grid seamstress, powers her vintage portable sewing machine by solar panel in North Carolina. Sewing machines need just an inverter to run off solar-charged batteries. A no-frills, affordable sewing machine designed for women in low-resource areas could provide them an additional income-generating stream, saving them time if they presently sew by hand. © Dan Kirby

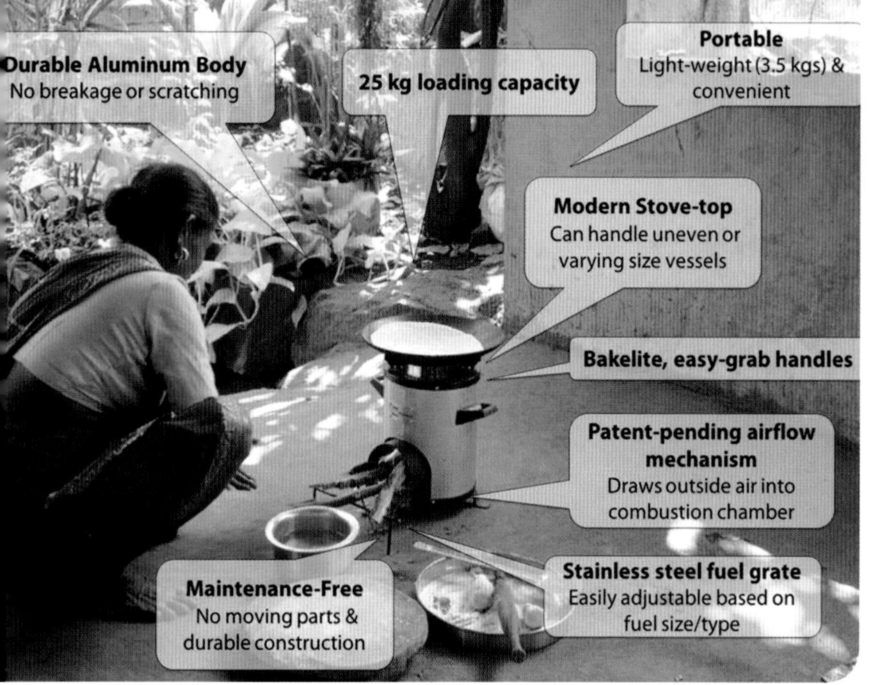

Greenway Grameen Stove's selling points, Rajasthan, India
© Greenway Grameen

Improved Cookstoves

Clean cookstoves consume less fuel, minimizing health-damaging smoke and black carbon emissions, and allow women to less time gathering wood.

H L&G tel

CleanCookStoves.org • @CookStoves
StoveTeam.org • @StoveTeam

Hundreds of millions of the world's poorest women still gather wood daily to fuel classic three-stone cooking fires. Open-fire cooking has many negatives:

- ✗ Women and girls spend many hours a day searching for and carrying wood.
- ✗ Smoke exposure endangers women and nearby children; an estimated 7 million people die annually from indoor air pollution.
- ✗ There is an ever-present threat of fires and burns.
- ✗ Direct burning of biomass generates **black carbon,** a major contributor to global warming.
- ✗ Deforestation lengthens the time spent seeking wood, and also increases soil erosion.

Adopting inexpensive **improved cookstoves**—also referred to as high-efficiency cookstoves—mitigates these perils. Hundreds of designs are on the market, tweaked for local fuel needs and cooking styles. Engineered to utilize fuel more efficiently, they offer a smorgasbord of financial, health, and environmental benefits:

- ✔ Lowered fuel consumption means families' expenses are lowered, paying for the cookstove and then saving money.
- ✔ Less time is required for wood gathering.
- ✔ Decreased smoke exposure improves women's and children's health and saves lives.
- ✔ Less food is burned, less time is spent stirring food to avoid burning, and food cooks faster.
- ✔ Children can safely be nearby during cooking without risk of burns.
- ✔ Less soot accumulates on pots, walls, surfaces, and fabrics, saving women work cleaning and scrubbing.
- ✔ Less black carbon is emitted, a global environmental benefit.
- ✔ Slowed deforestation provides more carbon absorption and stabilizes topsoil.

Despite the many reasons to upgrade to improved cookstoves, the transition has been frustratingly slow.

End users—women—were rarely consulted about their needs and preferences, so early designs were not aligned with their cooking methods. Women feared new stoves would compromise the quality or taste of their food.

- Designs often performed better in labs than in actual cooking venues.
- Many households use them as an additional stove, not a replacement.
- Stoves' designs must be compatible with available fuel sources.
- Improved cookstoves require a cash outlay beyond people's means. The poorer people are, the less willing they are to take risks on unproven, unfamiliar technologies.
- Though women benefit the most from improved cookstoves, men generally make the purchasing decisions and do not prioritize or desire improved cookstoves.

The Global Alliance for Clean Cookstoves, a United Nations Foundation-led effort to mobilize and coordinate resources, is addressing these challenges. Working with more than 900 partners, the Alliance seeks to overcome the barriers impeding the adoption of improved cookstoves and fuels. Promising trends include:

- ✔ Heightened attention to co-creating, testing, and evaluating cookstove designs in order to produce cookstoves women will be eager to use.
- ✔ Upgraded two-burner models.
- ✔ Flexible fuel designs accommodate a range of fuel sources.
- ✔ Vendors have launched microenterprises fabricating and selling fuel, creating a fuel supply chain. People only purchase a stove once, but need fuel daily, so this is smart enterprise.
- ✔ **Carbon offset subsidies** can help lower prices.
- ✔ Microcredit is well-suited to stoves; purchasers pay off their loans through savings in fuel outlays.
- ✔ Thirty-day guarantees and other customer-service promotions reassure wary buyers.
- ✔ Local demonstrations and happy users' testimonies speed adoption. Women are often hired as sales reps, speaking knowledgeably about their own cooking experiences to their friends, neighbors, and other prospective buyers.

Namigadde Mwamiin selling improved cookstoves and briquettes, Uganda. © courtesy of GVEP

Consolée Kavira, project manager, shows off a Black Power Stove. © Martin Wright/Ashden

"Village women thought cooking on a chula meant an open fire and smoke-filled kitchens. Now half a million women are cooking curry on improved stoves in smoke-free kitchens and enjoying it. Shakti has succeeded in changing village mindsets."
—*Muhammad Yunus*

Black Power Stoves, spearheaded by WWF-Democratic Republic of Congo, recently captured an Ashden sustainable energy award. Locally fabricated, stoves sell for $5–$10, paying for themselves in two weeks. Cooks use 50 percent less charcoal and achieve better results because the heat can be regulated.

A **Greenway Smart Stove**, (a Grameen initiative):

- ✓ Reduces fuel usage by 65 percent
- ✓ Reduces smoke emissions by 80 percent
- ✓ Costs around $23

Greenway CEO Neha Juneja, an industrial engineer who previously worked in air-flow systems design, applied these skills to designing the Greenway. Extensive consultation with local women paid off, as the Greenway is now the best-selling biomass stove in India.

"We traveled across five Indian states, living with multiple communities, and developed over ten designs with their inputs. Women who have been using mud stoves every day for years are extremely well-versed with the combustion principles of open fires—we were able to tap this knowledge to good use."
—*Neha Juneja*

Oregonian Nancy Hughes founded **StoveTeam** after serving on a medical mission in Guatemala and witnessing firsthand the terrible injuries women and children sustained from cooking fires. Their sturdy **Ecocina** stove can be fueled with small twigs, corncobs, coffee husks, or branches. Ecocinas:

"Day in day out, and for hours at a time, women and their small children breathe in amounts of smoke equivalent to consuming two packs of cigarettes per day."
—WHO Fuel For Life: Household Energy and Health

- ✓ Reduce carbon emissions by 68 percent.
- ✓ Reduce air pollution by 86 percent.
- ✓ Reduce wood consumption by more than 50 percent.

StoveTeam has facilitated the establishment of Ecocina factories in El Salvador, Guatemala, Honduras, and Mexico.

See also: **Eco-Briquettes (#32)**, **Green Charcoal (#33)**

- Sign on for a StoveTeam Central American **Stove Camp**, pairing volunteers and locals to fabricate and promote the Ecocina.
- Partner with Rotary and StoveTeam to open a new factory.

A Bolivian mother starts her solar box cooking in the morning. © E+CCO

Solar Box Stoves

Wooden box solar stoves accommodate most cooking needs and provide economic, health, and environmental benefits.

cedesol.org • @CedesolBolivia • SolarCookers.org • @SolarCookersInt • @She-inc.org • @SolarHousehold

The full potential of fuel-free solar cooking has not yet been realized, but refinements in solar stove designs, along with educating communities about their benefits, have increased uptake.

Solar panel cookers, cardboard with a reflective covering, are an easy do-it-yourself project.

Wooden box solar stoves feature inner reflective surfaces; a hinged top reflective frame; and a hinged, slanted, double-glazed glass top covering the box, where the food is placed. The bottom is painted black and the box is insulated, trapping more heat. Black pots further aid heat absorption. Temperatures reach between 120°C/248°F and 160°C/320°F, suitable for most cooking.

Solar cooking's benefits include:

- ✔ The elimination of fuel expenses.
- ✔ Eliminating the need to collect wood. This minimizes women's and girls' exposure to gender-based violence in conflict zones and refugee camps.
- ✔ Eliminates indoor cooking smoke.
- ✔ Saves time, since no fire needs tending, nor does food need stirring to prevent burning.
- ✔ Conserves forests and eliminates **black carbon** emissions.

Bolivia-based **Cedesol** and their partner, NGO **Sobra La Roca**, build eco-smart stoves from local materials. Purchasers can save money by building their own solar box stove at a Sobra La Roca workshop.

> "I use just a little firewood to get the food to boil, and then I place it into the solar stove. I have two to three hours to weave while the food cooks, increasing my income."
>
> —*Felicidad Orellana, Cochabamba, Bolivia*

Cedesol coordinates microfinancing for solar stoves, priced at around $75. Once they are paid off through savings on fuel, families have barely any fuel expenses.

VietnamSolarServe markets its design, the **SQ50 Box Cooker**, for around $50. Though Vietnam is blessed with 1,400–2000 hours of sunshine in the north and 2,000–3,000 hours in the central and south, few people know how to take advantage of it. Solar Serve is committed to outreach and solar cooking training.

Challenges:

- Sunlight is unreliable. It can be rainy or cloudy, or daylight hours may not line up with cooking needs.
- Advance planning and patience are required, since solar cooking is slower than conventional cooking.
- Solar cooking techniques may not align with local food preparation styles, requiring new cooking habits.
- Even where solar cookers are well-established, conventional stoves are still needed for backup on rainy or cloudy days, or when quick cooking is required.

YOU

- Cedesol offers certified **Gold Standard Carbon Offsets**, supporting education and low-income solar stove purchases for poor households.
- Cedesol offers internship opportunities. Requirements: a minimum two-month commitment and basic Spanish.

EDU

- A high-production-quality *Solar Box Stove Cooking Show* could help create demand. Consult Ruth and David Whitfield's *Cooking with Sol* and then produce a series.

Parabolic Solar Reflectors

Parabolic solar reflectors concentrate the sun's rays, bringing the contents of a suspended kettle or supported pot to a boil more quickly than other solar cookers.

L&G tel

Solarcooking.wikia.com • www.fost-nepal.org

Cooking dal-bhat (lentils and rice) in a parabolic solar cooker, Alapot, Nepal.
© Allart Ligtenberg, SolarCookingWikia

Parabolic solar cookers, also known as solar concentrators, are one of many technologies used to collect and absorb the sun's rays, along with solar panel cookers and box stoves (**#51**), SODIS/solar water disinfections (**#39**) and solar food dehydrators (**#68**).

Parabolic solar ovens are now a common sight in Nepal and Tibet, especially along high-altitude trekking routes, in no small measure due to Allart Ligtenberg's leadership in independently introducing solar technologies to Nepal in 1992. This beautiful, remote country shares the problems of **energy poverty** with many other regions. However, Nepal's high elevations and cold winters mean wood burning is also needed for heating, further stressing forests and causing respiratory disease.

Since retiring from his Hewlett Packard engineering job, Ligtenberg has involved Rotary in more than twenty Rotary Matching Grant initiatives, introducing, refining, and promoting solar tech in order to:

- ✔ Improve impoverished families' health by decreasing smoke exposure
- ✔ Utilize free sunlight, cutting households' fuel expenses
- ✔ Relieve women's and girls' double burdens of wood carrying and cooking over smoking fires
- ✔ Reduce deforestation

Parabolic solar collectors concentrate solar rays on a specific focal point:

- ✔ Heating contents more quickly than solar panel or box ovens
- ✔ Reaching higher temperatures than panels or boxes
- ✔ Functioning even in sub-zero weather (as do panel and box cookers)
- ✔ Used like a stove top for sautéing, stir-frying, browning, and boiling, pots or pans sit on the strategically placed shelf-rack
- ✔ The larger the size, the more food that can be cooked
- ✔ Pots need not be black, as is preferable with solar panel and box cookers

Users need to follow safety protocols and basic steps for cooking with parabolic collectors:

- Parabolic cookers can reach very high temperatures, even up to 450°, requiring caution.
- Unattended food can burn.
- Sunglasses are needed to protect from the glare of the sun's reflection; users should never look straight at the focal point.
- Pot lids should be used to avoid splattering the reflectors.
- Parabolic cookers' positions need to be adjusted to track the sun's movement.

Chinese manufacturers are producing parabolic reflectors in the $40 and up range.

Nepal's Katmandu-based **Foundation for Sustainable Technologies (FOST)** promotes parabolic solar reflectors, partnering with Allart Ligtenberg and supported by Rotary's efforts.

YOU

- **FOST** welcomes volunteers with a background in sustainable technologies.

EDU

- A discarded TV satellite dish can be repurposed as a parabolic reflector solar cooker. The most popular material for reflectors is mirror-finished anodized aluminum, available in sheets, or Mylar. These are cut into narrow triangular facets and riveted to the dish. Instructional videos and articles are available on the Internet.

A young Liberian checks out hot lunch prepared in a thermal retention basket. © Karyn Ellis/SolarCookers International

Thermal Retention Cooking

Thermal retention cookers are insulated containers holding pre-boiled pots of food. The contents continue cooking without additional fuel.

D·I·Y H L&G tel

solarcooking.wikia.com • @solarcookersint • Nb-wonderbag.com • @TheWonderbag

Heat retention cooking, a traditional fuel-conserving technique, utilizes an insulated box, basket, or bag. Food is brought to a boil, covered, tucked in the insulated holder, and closed up tight. Very little heat escapes, so the contents keep cooking.

They are known by various names, including:

- Fireless cookers
- Hayboxes
- Haybaskets
- Heat retention cookers
- Slow cookers
- Thermal cookers
- Wonder box cookers

Materials like straw, kapok, or hay are traditional, but whatever is on hand can be utilized, including old blankets or sleeping bags. Thermal retention cooking takes longer than conventional cooking, but provides many benefits for simmering stews, soups, porridges, and beans:

- ✓ Once the food reaches the boiling point, no additional fuel is consumed, reducing time spent collecting and carrying wood.
- ✓ Retention cookers can be fashioned from free or reused materials.
- ✓ Cooking requires no stirring or stoking the flame.
- ✓ Food does not burn, as often happens over a flame, meaning less wasted food and easier cleanup.
- ✓ These cookers use 25 percent less liquid, an advantage where treated water is scarce.
- ✓ Smoke-free, they improve air quality for everyone in the vicinity.
- ✓ Burn and fire hazards are eliminated.
- ✓ Soot-free, they save women time cleaning pots, walls, and surfaces.
- ✓ Less wood means less deforestation.

Do-it-yourself designs are posted at the **Solar Cookers Wiki**; many sewing blogs provide patterns for insulated thermal retention bags and boxes.

Malawi-style fireless cooker: a basket lined with an old blanket and banana leaves. © Betty Londergan/Heifer Fund

Thermal-retention cooking can be used in conjunction with other techniques for further energy conservation. **Improved cookstoves** (**#50**) cook more efficiently, using less fuel. SODIS (**#39**) heats water to nearly boiling from free sunlight.

Solar box ovens (**#51**) pair thermal retention cooking technology with enhanced solar energy capture.

Sarah Collins, an enterprising South African social innovator, created the **Wonderbag**, an appealing fireless cooker update. Floppy, styrene-insulated lidded bags are fashioned in bold fabrics and marketed using a buy-one, donate-one model. Microsoft has collaborated with Wonderbag to document fuel savings for WonderBag's **carbon offset funding**.

WonderBag Statistics – Quick Facts and Figures

- Carbon Tons offset per bag: 1 ton per bag, per year
- Disposable Income Created: $36.50 per bag, per year
- Trees Saved: 1.7 per bag, per year
- Water Saved: 156 liters per bag, per year
- Fuel Savings: 30 percent electricity, 60 percent charcoal/firewood
- Jobs Created: 1 job per every 1,000 bags

Heat retention is popular worldwide among those who want to use resources frugally and responsibly, regardless of their income or area's infrastructure level.

YOU EDU

- Make your own haybox and lower your cooking carbon footprint.

Improved Kitchen Equipment

Better-designed equipment saves women time and drudgery, improving their productivity and quality of life.

tel

LuckyIronFish.com • @LuckyIronFish • d-lab.mit.edu • @Dlab_MIT

An inexpensive maize sheller, twice as fast as hand-shelling. © One Acre Fund/Stephanie Hanson

Millions of women and girls spend hours each day carrying water and foraging for wood. After lugging their loads home, their next round of work begins. Food preparation in low-resource areas is very labor-intensive. Here are several pieces of equipment offering significant benefits at minimal cost.

ZhiLiang Wan's **Off-Fire Reboiling Pot**, based on thermal retention cooking principles (**#53**), is popular in China. After the rice (or any food cooked in liquid) boils, the burner is turned off; the covered food continues cooking. Around $15, they are only available in Asia. Introducing them worldwide would expand access to benefits:

- ✓ 50 percent reduction in fuel expense, quickly paying for itself
- ✓ Cooking food unattended
- ✓ Quicker cleanup time, since food doesn't burn
- ✓ Quantifiable reduction in carbon emissions, earning Clean Development Mechanism funding

Pressure cookers can be used over open flames to cook food in liquid. Popular in India, they cost around $20 and offer significant benefits:

- ✓ Lowered fuel outlays or time spent gathering and transporting wood, quickly paying for itself
- ✓ Speedier cooking time (unlike haybox or solar box oven cooking, which take more time than conventional cooking)
- ✓ No smoke or carbon emissions, promoting personal health and combating global warming

The introduction of nonstick cookware in the 1960s ranks high among labor-saving innovations. Mansukhbhai Prajapati, the Gujarati inventor of the **Mitti Fridge** (**#55**), also produces nonstick coated terracotta *tawas*—flat pans for preparing traditional foods like roti, parantha, and naan.

Combining the old and new at an affordable price point, Mitti Cool Tawas:

- ✓ Cost under $2, making them affordable for low-income households
- ✓ Use less oil and are easier to clean
- ✓ Consume 25 percent less fuel, quickly paying for themselves

The Cambodia-based social enterprise **Lucky Iron Fish™** markets an innovative iron supplement solution. Anemia is rampant in Asia in particular, causing a myriad of health problems (**#7**). Cast iron pots leech beneficial iron but are too expensive for the lowest-income population. Iron is similarly transmitted just by placing a lump of iron into cooking pots. By shaping the iron into a fish, a Cambodian good luck symbol, the company got locals to embrace the initiative, and users are showing markedly lower rates of anemia.

Hand-shelling corn, Ecuador. © Patty Grimm

MIT's famous **D-Lab** promotes their easily fabricated maize sheller hand tool, pictured above. Maize, a global staple, is left out in the field for kernels to harden. Women spend hours hand-shelling—slow and tedious work. Locally made metal shellers help do the job twice as fast, with less pain, for $2. Instructions are posted online.

See also: **Micronutrient-Fortified Food and Supplements** (**#7**), **Prenatal Supplements** (**#19**), **Bicycle Powered Maize Sheller** (**#30**)

YOU • Add more affordable, labor-saving tools to this list!

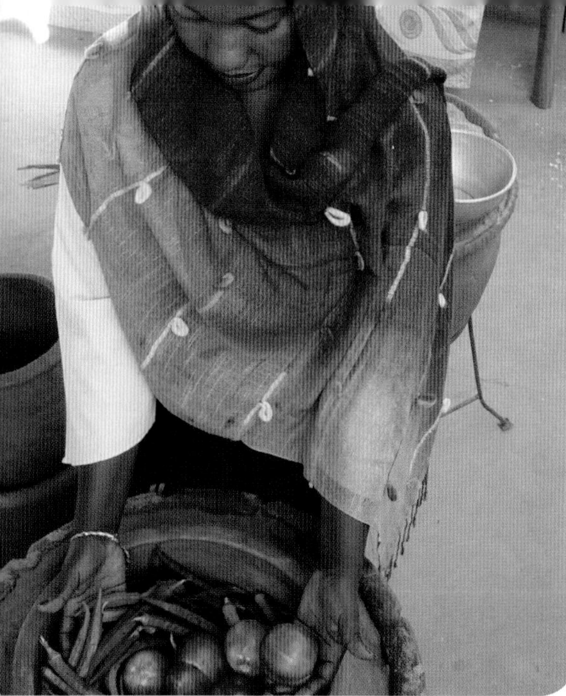

Okra and tomatoes stay fresh in a ceramic pot-in-pot, Sudan. © PracticalAction Sudan

Cold Food Storage

Evaporative coolers, made of ceramic and wet sand, keep produce fresh for many days, increasing farmers' and vendors' incomes.

PracticalAction.org • @PracticalAction

Refrigerators are absent from low-resource areas because there's no electricity, and most people can't afford to stockpile food. In hot climates, perishable foods quickly spoil, resulting in lots of wasted produce.

The **Zeer Pot**, invented by Mohammed Bah Abba in Nigeria in the 1990s, is a **Pot-in-Pot Cooling System** consisting of two porous earthenware pots, the smaller one set in the larger one with a layer of wet sand in between. Moisture evaporates and cools the inner pot housing produce or beverages covered with a damp cloth or ceramic top. Water is added twice daily.

> "Abba's first trials of the Pot-in-Pot proved successful. Eggplants, for example, stayed fresh for twenty-seven days instead of three, and tomatoes and peppers lasted for three weeks or more. African spinach, which usually spoils after a day, remained edible after twelve days in the Pot-in-Pot."
>
> – *Rolex Awards for Enterprise, 2000*

Evaporative cooling can help avoid food waste, though it doesn't replace refrigeration's steady 40° temperatures, safe for meat and dairy products. And they are not effective in humid climates. But for $2, they provide many benefits:

- ✔ Farmers and vendors (sometimes the same person, sometimes different individuals) can time their sales, rather than being forced to sell at low prices to avoid quick spoilage.
- ✔ Girls do not need to miss school to sell at local markets daily; vending can align with school schedules.
- ✔ Pot-in-pot production provides work for local potters.

The **Mitti-Cool Fridge** is the invention of Mansukhbhai Prajapati, a Gujarati ceramicist and entrepreneur who also manufacturers non-stick ***tawas*** (**#54**). Introduced in 2005, it is an upgraded evaporative cooler shaped like an antique ice box. The upper area houses a large water chamber outfitted with a tap, doubling as a water cooler. The water very slowly drips down the sides, promoting the evaporation that cools the inner shelf.

The evaporative cooling process utilized by the clay Mitti-Cool keeps vegetables and fruits fresh for six to seven days, while milk can be preserved for three days, a significant improvement over room temperature. As with the Pot-in-Pot cooler, the Mitti-Cool doesn't work if the climate is too humid. The Mitti-Cool, which is priced in the $90 range, uses no electricity, so has no operating costs.

The **ChotuKool**, an off-grid cooler, is another example of ***jugaad***, India's name for **frugal engineering** innovations. Only 20 percent of the population has refrigerators. Invented by Gopalan Sunderraman in 2010, it is the size of a picnic cooler and runs on a 12-volt battery, designed for both domestic and kiosk use—useful for beverage cooling and an ideal solution for selling chocolate.

EDU

- Instructions for creating Pot-in-Pot cooling systems are available online. Test them out, especially if you live in a warm, dry climate.

Laundry

Hand laundering is physically demanding and time-consuming. Scant attention has been paid to making this task less burdensome.

tel

SuSanA.org • @susana_org • susan-design.org

Lugging carpets on laundry day in Morocco © Margaret Shapiro

Without running water or electricity, laundering is done by hand, in open waterways or washbasins. Either dirty laundry needs to be hauled to the water, or the water needs to be lugged to the laundry basins. Garments are soaked in soapy water, scrubbed with a washboard or against a stone, rinsed, wrung out, and hung to dry—an entire day's work.

On a nice day, doing laundry at the waterfront with girlfriends may sound like a day at the beach, but the socializing benefit doesn't mean women wouldn't prefer faster, easier methods of laundering clothes. Plus,

- ✗ Women frequently suffer from repetitive stress injuries and skin irritation from hand laundering.
- ✗ The water may be contaminated, exposing launderers to pathogens.
- ✗ Dumping dirty, sudsy residue into open waterways or city streets is ecologically undesirable.

Village women often congregate centrally, washing their laundry in concrete tubs. In urban settings, water supplies are constrained and there is no mechanism for disposing of wastewater. As you'll note in the photo to the right, dirty water simply runs across the pavement.

Unfortunately, laundering generally falls under the radar of business, women's empowerment, and sanitation sectors.

Alex Cabunoc and Ji A You designed the **GiraDora** washing machine while enrolled in Pasadena's Art Center College of Design's **Design-Matters**. Following **co-creation** principles, they researched and then collaborated with low-income Peruvians, winning many design competitions with their foot-pedaled plastic drum, which is operated from a seated position. Legs are stronger than arms, making the job easier than hand-washing. Slated to sell for around $40, with no operating costs, the GiraDora:

- ✔ Uses one-third less water than hand laundering
- ✔ Leaves hands free (cell phones, anyone?)
- ✔ Extracts water like a salad spinner, cutting drying time
- ✔ Does an hour's worth of physical labor in five minutes
- ✔ Facilitates **greywater** reuse

The **GiraDora** has laundering microenterprise possibilities. The migration of single men from rural to urban areas creates a customer base for affordable laundry services. Though designed for low-resource areas, the GiraDora has generated a great deal of interest from potential industrialized world purchasers living in small quarters or intentionally off-grid. It is cheaper, more eco-friendly, and less time-consuming than heading to the laundromat.

Laundry day in urban India. © Water.org

A Philadelphia University industrial design student duo, Eliot Coven and Aaron Stathum, designed the **Up-Stream** laundry system. Their creation is assembled horizontally with a rope configured as pedals, constructed with under $20's worth of off-the-rack hardware. Production has stalled due to lack of funding.

Sector 6 – An Overview

Subsistence Farming

THE FARMER. Many people are surprised to learn that about half of the world's farmers, most leading a hardscrabble existence, are women. Increasingly, as men migrate to cities in search of employment, women and children are left at home in the village. Studies show that if female farmers had the same access to seeds, tools, credit, and technical training, their crop yields would increase by 20–30 percent.

THE SEEDS. Improved crop varieties produce higher yields and more nutritious crops (**#59**). Farmers are vulnerable to weather conditions, exacerbated by climate change. **Drought Resistant and Flood Resistant Crops** (**#58**) better withstand weather extremes.

THE SOIL. Expanding rural populations results in soil depletion due to overuse, and fields are not all created equal. If they're rockier, hillier, more eroded, with thinner soil—it's more likely they are farmed by the village's women. If women don't have **Land Titles** (**#98**), their fields are even more likely to be degraded.

Several of the entries in this section focus on improving farmland fertility through soil amendments like **Biochar** (**#57**), natural fertilizers, and soil-enriching crops. **Urban Farming** (**#66**) and **Keyhole Kitchen Gardens** (**#65**) have great potential as well.

THE TOOLS. Women farm with very primitive implements. Well-designed manual tools like **Wheeled Seeders** (**#61**) save hours of backbreaking labor, allowing women to reallocate that time to more productive efforts.

THE INPUTS. Crops need water, a job greatly eased by **Drip Irrigation** (**#64**). Affordable **Treadle Pumps** (**#63**) make accessing water easier, allowing dry-season crop cultivation and much higher incomes for farmers. **Urine Fertilizer** (**#62**) lowers overhead and helps produce bountiful crops.

THE LIVESTOCK. Raising **Bees** (**#70**), **Chickens** (**#71**), and **Goats** (**#72**) contributes to a farm family's nutrition and provides added income. **Fish Farming** (**#73**) also provides not only added income but also enhanced protein for hungry farming families living near waterways.

THE HARVEST. Improved yields are a breakthrough, but since significant percentages of harvests are lost to spoilage, farmers and everyone else lose out. Simple food preservation techniques like **Purdue Improved Crop Storage Bags** (**#67**) and **Solar Dehydration** (**#68**) mean increased food supplies and income.

THE EARTH. The planet's entire ecology is being transformed by human action: deforested woodlands, polluted waters, and degraded habitats and eco-systems. Repairing damage and mitigating global warming are large 21st-century challenges. Subsistence farmers can perform vital global environmental services through beekeeping, tree-planting, and biochar production and implementation. Compensating them for these duties is a win-win: it helps them escape extreme poverty while repairing our planet.

57. BIOCHAR SOIL AMENDMENT

Biochar enhances plant growth and helps soil retain water. It sequesters carbon, making it a climate-change mitigator.

58. CLIMATE-ADAPTED CROPS

Higher-yielding crops that better withstand flooding, drought, diseases, and pests improve food security and increase farmers' incomes.

59. NUTRITIONALLY ENRICHED CROPS

Boosting crops' nutritional content means farmers produce more health benefit per bite.

60. INTERCROPPING

Intercropping—sowing complementary plants in adjacent rows—improves soil, increases yields, and boosts income.

61. SEEDERS AND THRESHERS

Well-designed tools ease the hard physical labor of planting and threshing. Women's groups can purchase them jointly.

62. URINE FERTILIZER

Human urine is rich in nitrogen, the primary ingredient in fertilizer, and performs as well as or better than synthetic fertilizers.

63. TREADLE WATER PUMPS

Treadle-operated pumps lift water from wells or open water sources, facilitating crop irrigation, expanding yields, and improving food security.

64. DRIP MICRO-IRRIGATION

Drip irrigation delivers water efficiently. Low-cost systems liberate small landholders from hauling heavy water loads for hours every day.

65. KEYHOLE GARDENS

Keyhole gardens facilitate garden tending without bending. An embedded compost column channels food scraps and greywater, fertilizing the soil.

66. URBAN BAG GARDENING

Sack gardens utilize vertical space, useful for tight urban spaces. Local food supplies improve diets and provide livelihoods.

67. IMPROVED CROP STORAGE BAGS

Airtight triple bagging provides safe long-term storage for cowpeas and other crops.

68. SOLAR DEHYDRATORS AND NUT SHELLERS

Improved crop-processing techniques preserve yields and save farmers' time. Solar dehydrators and nut shellers are affordable when shared by women's cooperatives.

69. AGROFORESTRY: INTEGRATING TREES INTO SUBSISTENCE FARMING

Trees provide powerful ecological and economic benefits.

70. BEEKEEPING

Beekeeping is relatively inexpensive to set up, yields honey and byproducts, and offers local and global benefits.

71. CHICKENS AND EGGS

Poultry farming provides eggs and meat for sale, as well as for family consumption.

72. GOATS

Goats, traditionally raised by women, provide milk to improve family diet and income from selling milk and offspring.

73. FISH FARMING

Fish farming provides high-protein food for the family and surplus fish to sell.

Producing biochar in Sabana Grande, Nicaragua. © Vanessa Trevino

Biochar Soil Amendment

Biochar, a beneficial soil amendment, enhances plant growth and helps soil retain water. It sequesters carbon, making it an effective climate-change mitigator.

Biochar-International.org • @Biochar_IBI • CarbonRootsInternational.org • @CarbonRoots

57 Depleted, unproductive soil is one of the great challenges subsistence farmers face when eking out livelihoods from small landholdings. Soil productivity is diminished by many factors, including:

- ✗ Failing to allow fields to lie fallow for a season (time off for the earth to regenerate)
- ✗ Erosion and general eco-system damage, often caused by deforestation
- ✗ Overuse of synthetic fertilizers, which, ironically, strip soil of minerals and beneficial microbes
- ✗ Monocropping deprives fields of the ecological benefits of biodiversity

Biochar, produced by carbonizing plant material in an oxygen-free chamber, helps restore soil's productivity. Using charcoal to enhance soil fertility is an ancient technique mastered by indigenous people in the Amazon Basin who built up rich *Terra Preta* ("dark earth") soil pockets; it has also long been utilized in Japan.

Biochar production is low-tech and can be produced on a small, local scale. Online, many **open-source** biochar production designs are available. A number of improved cookstove models burn crop residue for fuel and produce biochar as a valuable byproduct. Carbonized plant material easily crumbles into powder.

Biochar powder is the basic ingredient of **EcoFuelAfrica**'s (**#33**) fuel briquettes. They source biochar from local farmers, who produce it from crop residue. The farmers retain a portion of each batch to improve their own farmlands.

Biochar benefits include:

- ✔ The utilization of locally available waste materials
- ✔ Decreased reliance on more expensive synthetic inputs, due to improved soil fertility
- ✔ Improvement in soil structure, which increases water retention and upgrades microorganism habitats
- ✔ Sequestering carbon—when crop waste is burned or left to decompose, it releases carbon into the atmosphere, contributing to global warming; biochar production carbonizes crop waste, transforming it into a solid carbon material that can be returned to the soil, which prevents the carbon from releasing into the atmosphere
- ✔ Once added to the soil, biochar remains intact indefinitely, so it is a long-term investment in soil improvement; farmers can keep on upgrading their soil by adding biochar

Biochar's carbon sequestration properties provide a monetizable global benefit. Paying smallholder farmers to provide ecological service would be a win-win.

As biochar's benefits become more widely known and better quantified, it has the potential to become an income-generating activity, not just a soil improver. The recently launched **Carbon-Roots Haiti**, under the auspices of Carbon Roots International, produces and sells biochar as an agricultural product.

Biochar is a soil amendment that upcycles waste, sequesters carbon, and improves soil structure, but it is not a fertilizer. Biochar International suggests mixing it with compost, which can provide the fertilizer, for gardening.

YOU
- **Re-char.com** is a social enterprise with a business model for low-income farmers to produce and market biochar. Perhaps you can replicate it.

- **Biochar-International.org** features many award-winning biochar science fair projects.

Climate-Adapted Crops

Higher-yielding crops that better withstand flooding, drought, diseases, and pests improve food security and increase farmers' incomes.

H

avrdc.org • @go_vegetables • CIMMYT.org • @CIMMYT

Saving maize seed in Dumka, India. © PhotoShare/Somenath Mukhopadhyay

Farmers contend with many variables: weather, crop choices, where to sow what seeds. Families have little room to experiment when their entire survival depends on the results. Climate change is causing more extreme weather events and even more unpredictable seasonal patterns.

Plants that can thrive despite predictable challenges like droughts, floods, and pests attain higher yields. A bigger harvest for the same amount of labor increases income and available food for a farmer, her family, and her community.

The ancient practice of breeding plants with success-enhancing traits is now informed by modern scientific methods. Preserving indigenous varieties of crops is the cornerstone, combined with meticulous laboratory experimentation, documentation, and pilot field-testing.

The International Maize and Wheat Improvement Center (CIMMYT), based in Mexico, focuses on these two global mainstays. CIMMYT maintains an **open-source** maize and wheat seed bank with more than 175,000 varieties of plants.

Work on drought-tolerant maize is paramount. For example, CIMMYT introduced new maize varieties in Malawi, ZM 309 and ZM 523, which counteract erratic weather patterns and frequent droughts, featuring:

- ✔ Earlier maturation
- ✔ Higher yields
- ✔ Better resistance to common leaf diseases
- ✔ Kernels that are easier to pound into flour, saving labor
- ✔ Open pollination, so farmers can save seeds and replant the following season, a significant savings

Locally, ZM 309 is known as *Msunga banja*, "that which takes care of or feeds the family," while ZM 523 is called *Mwayi*, meaning "fortunate."

Rice is the main dietary staple for almost half of the Earth's population. Grown in paddies, flooding is a major threat—young rice plants can generally survive only three days underwater. Dr. Pamela Ronald heads her own laboratory in Davis, California, which has developed rice that can survive submerged for ten days.

> "When I was in India . . . I met with about twenty rice farmers who had recently switched to a new rice seed called Swarna-Sub1, which is both very productive and can survive in flooded fields. Their rice fields get flooded every three to four years, and in past flood years, they ended up with almost no food to eat. Now, these farmers can feed their families no matter the weather . . . . as farmers in the region adopt Swarna-Sub1, they will grow enough extra rice to feed 30 million people."
>
> —*Bill Gates, 2012 Annual Letter*

Plant breeders select for a variety of additional beneficial traits, including:

- ✔ Resistance to diseases and pests
- ✔ Quicker maturation
- ✔ Staggered maturation, so farmers can likewise stagger labor and sales/income
- ✔ Crops that are easier to harvest and process, without loss of food quality
- ✔ Faster-cooking crops that save fuel

Once new strains are ready for dissemination, the next challenge is creating distribution supply chains that reach farmers in remote areas.

YOU

- **AVRDC,** The World Vegetable Center in Taiwan, has internship opportunities.

A technician uses traditional plant breeding methods to develop vitamin A-rich cassava, Nigeria. © Mel Oluoch/HarvestPlus

Nutritionally Enriched Crops

Boosting crops' nutritional content means farmers produce more health benefit per bite.

cipotato.org • @CIpotato • HarvestPlus • @cgiar.org • @CGIAR

59 Around a billion people in the world are malnourished. They don't consume enough calories, and their diet is micronutrient-deficient. Malnourished children are at high risk for maladies that can significantly impede their development and lower their disease resistance.

Biofortification, breeding crops rich in the micronutrients most lacking—iron, zinc, and vitamin A—is a cost-effective approach to improving nutrition.

- ✔ People derive the benefits while eating familiar foods, without behavioral changes.
- ✔ Once biofortified crops are established, seeds are saved for the next crop, a repeating cycle offering a high rate of return on initial research and distribution.
- ✔ Biofortification is a local solution; fortified staples or supplements are not always accessible for rural populations.

If biofortified strains are high-yielding, as well as drought- and pest-resistant, it greatly multiplies their value.

Sweet potatoes are excellent at micronutrient delivery. One of the world's most widely cultivated plants, they have spread far beyond their native Latin America. **HarvestPlus** has introduced a biofortified, orange-fleshed sweet potato high in beta-carotene, which the body converts into vitamin A. It is drought-resistant and higher yielding. Tests show that children who consume them have less vitamin A deficiency. Sweet potatoes have added virtues, since they:

- ✔ Produce more edible calories per hectare per day than wheat, rice, or cassava
- ✔ Mature in three to four months, faster than white and yellow potatoes
- ✔ Grow in marginal land and require little farmer effort
- ✔ Need very little chemical inputs like pesticide and/or fertilizer
- ✔ Have edible leaves and vines, in addition to edible roots
- ✔ Provide nutritious, cheap fodder for livestock

> **Animals fed on sweet potato vines emit less methane gas, important for mitigating climate change.**

The Sweetpotato Action for Security and Health in Africa (SASHA), a program of the Peru-based **International Potato Center**, is dedicated to maximizing worldwide uptake of roots and tubers to alleviate poverty and provide food security. They are especially attuned to reaching women. One innovative approach, Mama SASHA, provides free sweet potato vines at prenatal check-ups. This gift improves prenatal health, and also upgrades recipients' older children's diet.

Harvest Plus is also working on biofortifying cassava and maize, global staples, with provitamin A.

To combat iron deficiency, HarvestPlus is introducing biofortified, iron-rich pearl millet, targeted for India, and iron beans, focused in Africa.

Zinc deficiency is prevalent among malnourished people, lowering disease resistance and causing childhood stunting. Zinc-fortified rice has the potential to reach the half-billion malnourished people who survive on rice-based diets.

Golden Rice is genetically modified to include beta-carotene, from which the body produces vitamin A. It will be distributed at no cost to subsistence farmers, improving their health, though its genetically modified status has aroused controversy.

See also: **Micronutrient-Fortified Food and Supplements (#7)**

Intercropping

A farmer pointing to her high yields from intercropping maize and gliricidia trees, Malawi. © World Agroforestry Center/ICRAF

Intercropping — sowing complementary plants in adjacent rows — improves soil, increase yields, and boosts income.

WorldAgroforestry.org • @ICRAF • SemillaNueva.org • @semillanueva

Depleted soils are a poverty trap: they demand more labor but produce less. Women are generally relegated to the least productive farmland, with less access to inputs, tools, credit, and agricultural education, locking them into low yields.

One effective practice to boost yields is **intercropping.** A classic version is the Native American **three sisters** companion planting of maize, corn, and beans.

Likewise, alternating trees and maize rows increases maize harvests. The trees, of course, only need to be planted once and grow indefinitely; maize is sown for each crop.

A twelve-year study in Zambia and Malawi conducted by the Nairobi-based **World Agroforestry Center** demonstrates that intercropping leguminous trees like *gliricidia* or *faidherbia albida* with maize boosts yields by as much as 50 percent. The intercropped field was compared to maize-only fields, one utilizing synthetic fertilizer and the other with no inputs. The nonfertilized, intercropped field bested the fertilized. The status quo for subsistence farmers, maize-only with no fertilizer, came in last.

"Fertilizer trees" provide many benefits:

- ✔ Absorbing nitrogen from the air and fixing it in the soil, lowering or eliminating the need for costly fertilizer
- ✔ Shedding leaves in the rainy season, they do not compete with crops
- ✔ Dropping leaves that return high-nitrogen organic matter to the soil
- ✔ Improving soil's water retention and protecting against erosion
- ✔ Supplying edible, high-protein leaves for livestock feed
- ✔ Providing habitat for beneficial birds and bees

Guatemala-based **Semilla Nueva** ("new seed") focuses on:

- Boosting yield through sustainable agricultural techniques
- Farmer-to-farmer education

A farmer who has been successful using new techniques becomes an enthusiastic, credible advocate who shares her new expertise with her fellow farmers—generally more effective than top-down advice.

Children in the rural communities where Semilla Nueva works have up to a 79 percent rate of stunting due to insufficient protein and micronutrients. Semilla Nueva works with local women to introduce nutritionally superior foods like high-protein pigeonpeas.

Semilla Nueva communities are now growing pigeonpeas, also successfully introduced in East Africa. Planted between maize rows, pigeonpeas provide environmental, economic, and health benefits:

- ✔ Drought resistance
- ✔ Cutting fertilizer costs by fixing organic nitrogen
- ✔ Decreasing soil compaction
- ✔ Open-pollinating, allowing farmers to save seeds
- ✔ High-protein, nutrient-rich food
- ✔ Yielding enough to feed families and still produce surplus crop to sell

Semilla Nueva is working to connect with the global pigeonpea supply chain. Networking with international buyers and meeting their standards helps subsistence farmers better target their operations.

Several Idaho Rotary Chapters, along with other Rotary groups, support Semilla Nueva.

See also: **Trees (#69)**, **Fair Trade (#93)**

- **Semilla Nueva** has a fellowship program and also welcomes interns on a project-by-project basis.
- **Semilla Nueva** also offers Immersion Work Trips; voluntourists work side by side with Semilla Nueva farmers.

An El Salvadoran farmer is eager to try out a wheeled seeder in Rancho Grande, San Vincente. © Chuck Haren/Plenty International

Seeders and Threshers

Well-designed tools ease the hard physical labor of planting and threshing. Women's groups can purchase them jointly.

csisa.org • @CIMMYT.org • @CIMMYT • Plenty.org • @IFAD.org • @IFADnews

61 Women's co-ops—also known as collectives, associations, and self-help groups—give female farmers strength in numbers. In traditional societies where women are rarely educated or allowed public roles, women's groups are forums where women can:

- ✔ Develop leadership and communication skills
- ✔ Upgrade to better tools, affordable when shared
- ✔ Gain better access to valuable farming information
- ✔ Self-advocate and connect to outside services like microfinancing, education/literacy initiatives, and health providers

Cooperatives introduce innovations that result in all participants improving their yields. Solo farmers cannot afford tools like wheeled garden seeders, but a typical women's self-help group, usually comprised of ten to fifteen women, can buy one. Members takes turns using it, and it can also be rented to nonmembers, generating income. An enormous amount of labor is saved; members now have newly found time to devote to other activities and pursuits.

A **wheeled garden seeder** is designed with interchangeable plates that accommodate different seed types. Planting is much more rapid, efficient, and precise. The Boks-FunnyFarm blog describes the process: "This little baby opens the furrow, drops the seed in at predetermined spacing, pulls the soil over the seed, and firms the soil."

An additional benefit is that farmers' back strain is reduced by standing upright rather than bending over to plant each seed. Seeders retail for around $100 and up. Presumably local manufacturing and bulk production could lower prices.

Another labor-saving device is a **paddy thresher**, using a foot pedal to separate grains of rice from the stalks and husks:

> "Female farmers in the Dihuli Village in the Muzaffarpur district of Bihar, India, participating in CSISA were provided a paddy thresher that eases their hardship in terms of labor and decreases their pain in terms of health. Work that used to take days now takes hours with the use of the tool."
>
> —*Madhulika Singh, CSISA Agriculture Specialist*

> **"Poor farm women not only work longer hours than men, but often perform physically demanding work."**
>
> ***–IFAD.org***

Cultures shape attitudes and beliefs about appropriate agricultural tools. For example, studies by the International Fund for Agricultural Development, IFAD, revealed fierce attachments to hoe designs in different African regions.

Pushing a wheeled seeder saves this farmer's time in Rajapur, Bihar, India. © Madhulika Singh/CIMMYT, CSISA

- Longer-handled hoes would be far easier on women's backs, but farmers who stand up straight are considered lazy.
- Though females are gradually performing more and more farming duties in Africa, tool designs have not been adapted to better serve women's needs.
- Men generally purchase tools for their wives, reflecting women's lack of agency.

Urine Fertilizer

Human urine is rich in nitrogen, fertilizer's primary ingredient, and performs as well as or better than synthetic fertilizer. For free. For real.

 tel

SuSanA.org • @susana_org • susan-design.org

Leafy greens thrive on urine fertilizer at the Chisungu School's garden in Epworth, Zimbabwe. © Peter Morgan/SuSanA

Human urine is rich in nitrogen, potassium, and phosphorous, exactly the macronutrients plants need for growing. Gardeners have long been aware of pee-cycling as a frugal, eco-friendly practice, but urine is now being taken seriously as an agricultural input for subsistence farmers. Why not utilize this free asset, especially when most subsistence farmers cannot afford to buy synthetic fertilizer?

Urine excreted by healthy individuals is generally benign and, if separated from human feces, presents little risk of spreading pathogens. In low-resource regions urine is collected by urine-diverting latrines or simply by providing buckets. Those accustomed to indoor plumbing may find this practice off-putting, but flushing urine wastes valuable nutrients.

- On average, adults produce around 500 liters of urine per year, yielding
  - 2.18 kg of nitrogen
  - 0.87 kg of potassium
  - 0.2 kg of phosphorous
- Fertilizer urine is stored in closed containers for at minimum a month to ensure it is uncontaminated. Urine produces ammonia, which kills all potential pathogens.
- Urine can be used at full concentration, though some recommend diluting with water.

Dr. Sridevi Govindaraj received her doctorate in ecological sanitation from the Bangalore University of Agricultural Sciences and has overseen studies verifying diluted urine fertilizer's impact on maize, bananas, radishes, tomatoes, millet, and French beans. In nearly all cases, urine outperformed chemical fertilizer.

In Philippine trials, crops grown with urine fertilizer yielded 1.5 to five times more than plots where no urine was used.

Urine can improve soil when used in compost:

> "Another alternative urine use is to add urine as a nutrient source in compost production. While the direct use of urine as a liquid fertilizer only mimics conventional agricultural practices by adding mere mineral nutrients to the plants, the production of urine-enriched compost offers a way of improving the soil condition as a whole."
>
> —*Gensch, Miso, and Itchon* *(see reference below)*

The Norwegian NGO **Design without Borders** has refined a $3 unisex **ecological domestic urine collector**, an effort headed by Uganda-based Sarah Keller, that:

- ✔ Incentivizes urine collection
- ✔ Provides privacy, prized by women who participated in the design process
- ✔ Minimizes the smell caused when urine breaks down and forms ammonia

For more info, contact Sustainable Sanitation Design—**www.susan-design.org**

YOU
- Practitioners (and curious gardeners) seeking an implementation guide: Download Sustainable Sanitation Alliance's publication, *Urine as Liquid Fertiliser in Agricultural Production in the Philippines*, by Gensch, Miso, and Itchon.

EDU
- Seeing is believing. School and demonstration gardens are excellent venues to test and spread new ideas. A slide presentation chronicling urine-fertilized maize growth at the Chisungu School garden, Epworth, Zimbabwe (pictured above) is available at Susana.org. Results: Their urine-fertilized maize grew thirty times larger than unfertilized maize.

Young girls pump using a treadle, in India. © Sarah Butler-Sloss/Ashden.org

Treadle Water Pumps

Treadle-operated pumps lift water from wells or open water sources, facilitating crop irrigation.

tel

kickstart.org • @KickStart_Intl
Appropedia.org • @appropedia

Treadle pumps replace hand-operated pumps for pulling water for irrigation (**#64**) up to ground level from:

- Boreholes
- Wells
- Lakes, rivers, and streams

Foot-operated treadle pumps were developed in Bangladesh by IDE in the 1980s and are widely popular. Promoted and sold by NGOs and social enterprises, they are generally financed through microloans (**#88**) paid off when farmers are flush from selling increased yields.

Kenya-based **KickStart International** has an impressive record, selling nearly a quarter million units. Their $30 **Money Maker Hip Pump** is a lightweight, female-friendly design. KickStart's Mobile Layaway, a secure payment plan, lets customers determine their own schedule and remit via cell phone.

Treadle pumps are easily operated by women and children. Treadles:

- ✔ Lift water six times faster than traditional hand pumps.
- ✔ Are powered by leg muscles rather than arms, giving operators more force and endurance.
- ✔ Have no fuel costs for operation.
- ✔ Save about half a ton of CO_2 each year, if they replace diesel pumps.
- ✔ Can be dual-operated by partners working together, increasing output.
- ✔ Can be locally manufactured, pumping resources into the local economy.

Treadle-pump designs continue to be refined. Suction treadle pumps pull up water rather like drinking straws, well-suited for lifting water from shallow waters in places like Bangladesh. Pressure treadle pumps were designed for use in Africa, where water is deeper and needs to be pulled up further, pushing it out and spraying it above the pump level.

Treadle pumps make it possible for farmers to access irrigation's benefits:

- ✔ Expanding the quantity of cultivated field land, and/or growing thirstier, higher-value crops like kale, cabbage, and tomatoes
- ✔ Increasing yields and speeding maturation through optimized watering
- ✔ Growing a wider variety of crops, diversifying income
- ✔ Adding one, or even two, dry-season crop cycles
- ✔ Increasing household income
- ✔ Improving farming families' diets by increasing food quantity and variety

Challenges:

- Though photographs feature smiling girls and women pedaling, it is hard work, performed outdoors in the sun. An average irrigation watering takes three hours. When incomes grow, women sometimes hire young men to do the pumping.
- Water is a finite resource. Africa has deployed only a small amount of its water resources for irrigation, but profligate irrigation has lowered water tables in some industrialized regions, as well as in India. Pairing treadle pumps with hyper-efficient drip micro-irrigation is an important water conservation strategy.
- Treadle pumps allow farmers to expand their cultivation capacity, but population pressures sometimes preclude accessing more land.
- Treadle pumps require a secure water source in or near agricultural fields. Treadles lift water. Unfortunately, they don't provide the water or move it laterally over long distances. Women and girls do that, often traveling many miles to the closest source of water and carrying it home.

- Kickstart posts internships.

Drip Micro-Irrigation

Drip irrigation delivers water efficiently. Low-cost systems liberate small landholders from hauling heavy water loads.

IDE.org • @IDEorg

Gravity-fed drip micro-irrigation in Nepal. © IDE

Manual irrigation is labor-intensive, heavy work. It is also inefficient, since much of the water evaporates off the surface or becomes runoff. **Drip irrigation**, an Israeli innovation, is the targeted delivery of water fed by a central water reservoir to plant roots along water distribution lines. It is renowned for producing "more crop per drop."

While it was a major breakthrough for commercial farming in water-scarce regions, drip irrigation remained unaffordable for small-plot farmers. The introduction of lighter-weight, affordable drip micro-irrigation systems, powered by gravity, brings drip irrigation's benefits to subsistence farmers.

Paul Polak, founder of **International Development Enterprises (IDE)**, champions business solutions to alleviate poverty; he is a huge proponent of de-engineered drip micro-irrigation systems. When attached to a water source such as rainwater baskets (**#35**) or treadle pumps (**#63**), small-plot drip irrigation systems expand subsistence farmers' opportunities.

- ✔ Micro-drip provides insurance**.** Storing water and using it for irrigation offers protection from droughts and late rains. The ability to irrigate and produce steady incomes decreases subsistence farmers' risks and lessens the need for men to migrate to cities for off-season work.
- ✔ Micro-drip saves labor**.** The gravity-fed drip irrigation system developed and sold in Myanmar by **Proximity Designs** works with water sources raised as little as three feet above ground level. It mechanizes a process formerly done by farmers carrying water over their shoulders, hour after hour, hand-watering plants. The $33 system cuts water use by 50 percent and raises yield by 33 percent. Micro-drip is well-suited for women to manage, and they can use the time they save more productively. Automated irrigation promotes children's school attendance.

> "Watering with the sprinkler cans every day, my life would be a short one. This entire burden was cut by drip. You can't measure that with money; it's invaluable."
>
> —*Proximity customer Ko Myo Myint*

- ✔ Micro-drip adds an extra crop cycle. In his book *Out of Poverty,* Paul Polak writes about the extra crop-growing season made possible by drip micro-irrigation. Utilizing stored rainwater, small-plot farmers irrigate and grow vegetables during the dry season. The beauty of this enterprise is that the produce is brought to market when it is most sought after and fetches the highest prices.
- ✔ Proceeds from this bonus harvest generally pay off the system and generate positive cash flow within a single season. Even inexpensive systems last a few years, so they can be used for several subsequent seasons with little or no equipment replacement costs.
- ✔ Micro-drip improves diet. Blemished dry-season vegetables unsuitable for the market provide a healthier, more varied diet for the farmer and her family.
- ✔ Micro-drip conserves water. Many low-income farmers utilize inefficient flood irrigation, stressing water tables and making water more difficult to access. Drip irrigation uses water far more frugally while improving yields, a win-win.

See also: **Treadle Water Pumps (#64)**

Maabisi Phooko tends her keyhole garden in Thaba-Sephara, Lesotho. © Kim Pozniak/Catholic Relief Services

Keyhole Gardens

Keyhole gardens, built from free materials, facilitate garden tending without bending. An embedded compost column channels food scraps and greywater, enriching the soil.

SendACow.org.uk • @SendACow • AGoodFoundation.ca

Keyhole gardens are sited adjacent to homes, allowing families to grow vegetables for eating and also to sell. They integrate raised vegetable beds and composting, two popular gardening techniques. A frugal innovation created from free, local materials, the gardens utilize food scraps and recycled **greywater** to nurture plant growth.

Raising the walls of the bed to waist height and embedding a composting system have popularized keyhole gardens for their:

- ✔ High yields, feeding a family of eight and still producing a surplus to sell
- ✔ Extension of the growing season (their stone walls retain heat, expanding cold climates' growing season at both ends); sometimes canopies are erected to protect the plants
- ✔ Convenient access, allowing gardeners to work standing up
- ✔ Compost simplicity—a compost column is built into the keyhole's center, making it easy to toss in scraps and wastewater and eliminating the need to:
  - Transport waste to a separate compost pile
  - Tend the compost pile
  - Deliver finished compost to the garden site and integrate it into the soil
- ✔ Reduced need for artificial fertilizer (keyhole gardens self-fertilize)
- ✔ Low-maintenance requirements; little weeding (because the vegetables are planted very densely) and little additional watering beyond **greywater** are necessary
- ✔ Flood protection, being high above ground level.
- ✔ Aesthetic appeal, with their stone walls imparting a sense of dignity and importance.

Humanitarian aid workers influenced by permaculture gardening principles introduced keyhole gardening in Lesotho to combat chronic food insecurity. Keyhole gardening addresses many of Lesotho's significant challenges:

- A high-altitude, harsh climate
- High rates of poverty and malnutrition
- Many children orphaned by the HIV/AIDS pandemic being raised by grandmothers

The gardens provide year-round food easily grown by elders. Producing large yields of up to five varieties of vegetables, the gardens promote health and expand dietary diversity.

> "Neighboring villages outside the project intervention area were reproducing keyhole gardens on their own initiative, clearly indicating the success of the intervention and its potential sustainability."
>
> —*TECA/FAO*

A Good Foundation, founded by Canadian Paula Washington when she was just eighteen years old, has built dozens of keyhole gardens through their Lesotho **GROW** initiative.

Keyhole gardens are circular, with a pie-piece cut-out allowing direct access to the center compost column. Viewed from above, they resemble the shape of its namesake, an old-fashioned keyhole.

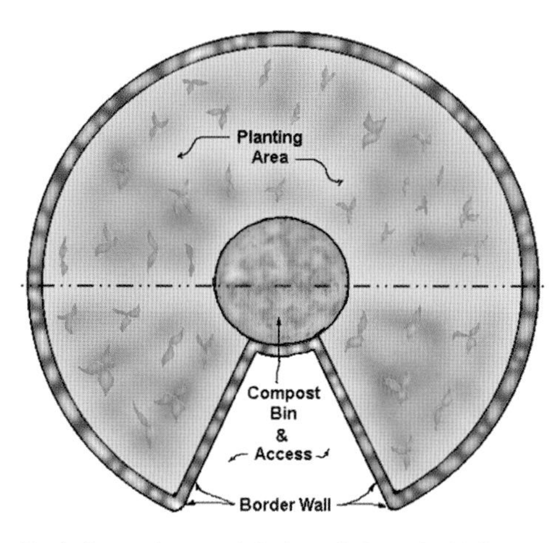

Keyhole garden, aerial view © Beverly Walker

- The width of the garden is only two yards across, allowing gardeners to reach plants from the center or the perimeter.

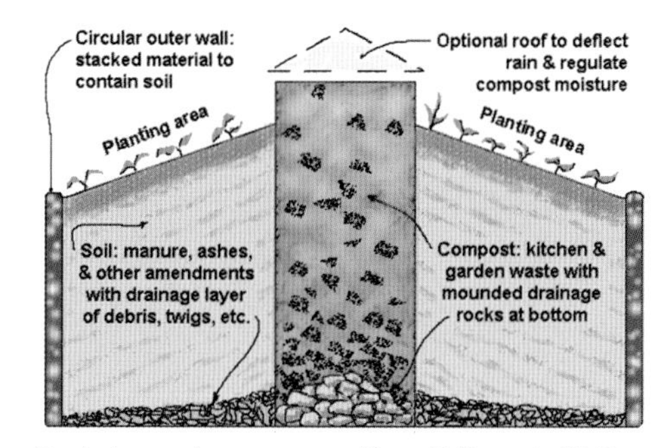

Keyhole garden cross-section © Beverly Walker

- Keyholes feature a central vertical compost basket column held in place by four poles.

Before: Lerato Thakholi dumps greywater into the keyhole garden's compost column, supported by four posts, in Lesotho. © A Good Foundation

After: Three months later, the Lesotho keyhole garden is thriving. © A Good Foundation

- The fill is a flexible mix of manure, hay, straw, leaves, ash, tin cans (which leech beneficial iron into the mixture), bones, wastepaper, and dirt, gradually transforming into rich, productive soil.
- The top is slanted downward from the compost column, increasing planting surface area and allowing excess water to run off.
- A lid placed over the compost column retains heat, reduces evaporation, and speeds composting.
- Pee-cycling (**#62**), if used, adds beneficial nutrients.

"Crop rotation and growing of insect-repellent plants are important to balance nutrient demands, fight insects and plant diseases, and deter weeds... A portion of the garden should be left fallow during each growing season."
–TECA.FAO.org

Keyhole garden costs are hard to pinpoint. While labor-intensive, the construction can often be provided by the owner. The materials are freely available in rural areas. The main cash outlay is for seeds.

In 2008, the BBC ran a story featuring Lesothos's keyhole gardens. The idea took off around the world, especially in the Southern United States—then experiencing a multiyear, extreme drought—because they:

✔ Retain water effectively
✔ Filter **greywater**, a frugal water reuse technique
✔ Are beautiful, flexible constructions

Southern gardeners build keyholes out of all manner of materials, from a junked rowboat to empty wine bottles. The results are the same wherever they are: beautiful, functional, water-sipping, high-yielding kitchen gardens that are more easily tended than conventional vegetable plots and also offer easy, onsite composting.

Oklahoma keyhole gardener Freddy Hill warns against siting a keyhole garden near a tree. Its roots will siphon the garden's water supply.

Mel Bartholomew's **Square Foot Gardening** technique of high-density-style vegetable plotting, popular in the United States, aligns well with keyhole gardening.

YOU

- There is no organization devoted to promoting keyhole gardening. Start one to:
  - Disseminate information
  - Share resources
  - Test alternatives and quantify results
  - Organize work-travel trips
- Build your own keyhole garden. Instructions are posted at Send A Cow.org and elsewhere. Building bottle brick (**#74**) retaining walls would be even more eco-friendly.

YOU EDU

- Partner with a school, church/synagogue/mosque, or community garden to create a demonstration keyhole garden, educating people about sustainable techniques for feeding the world.

A container gardener/farmer in Kibera, Africa's largest slum.
© Eric Wamanji

Urban Bag Gardening

Sack gardens utilize vertical space, useful for tight urban spaces. Locally grown food improves diets and provides livelihoods.

solidarités.org • @Solidarites_Int • Appropedia.org • @appropedia

The world's population is trending urban. Most of the world's twenty-two megacities—areas with over ten million people—are in the Global South. Sprawling, densely-packed informal settlements, haphazardly constructed and lacking basic infrastructures, are magnets for impoverished rural migrants. Available work is usually informal, irregular, and low-paying.

Sack gardening was introduced in 2008 in Kibera, just such a slum, by **Solidarités International**. It proved popular, providing nutritious food, as well as meaningful work, for this impoverished, underemployed population.Growing one's own food is empowering and taps rural migrants' agricultural skills, generating income from selling their surplus.

Often food aid bags are reused and filled with soil. Seedlings are planted in side slits, strawberry pot–style. Tubers are planted in the bag's top plane. Sack gardening offers micro-farmers many advantages:

- ✔ No transportation footprint, lowering food costs; customers are right there
- ✔ Provides a hedge against rising food prices
- ✔ Portability, significant for squatters and renters who lack legal titles to the land on which they garden

Sack gardening has been replicated successfully in other venues. An FAO-Netherlands Partnership Programme introduced them in Gaza, estimating ten garden sacks could feed a family. Bag gardens:

- ✔ Don't require plowing, much weeding, or heavy labor (though the sacks are heavy)
- ✔ Extend the growing season
- ✔ Have modest water requirements; a central column of small stones draws water throughout the whole sack
- ✔ In school settings, supplement students' diets; kids learn to grow their own food

Innovations described in other entries are relevant for sack gardening:

- ✔ Small-scale rainwater harvesting (**#35**) decreases dependence on urban water vendors.
- ✔ Urine fertilizer (**#62**), easily collected in high-density neighborhoods, boosts yields.
- ✔ Sacks can be connected to a drip micro-irrigation system (**#64**). One full water barrel feeds the setup, automatically dripping water throughout the day.

> **Common crops include tomatoes, onions, spinach, kale, spider plant, squash, amaranth, fodder, and African nightshade, with kale currently enjoying the most success.**
>
> *–Appropedia.org*

Challenges:

- Soil + compost + manure + seedlings must be obtained.
- Seeds cannot be planted directly in the bags. Seedlings must be transplanted into the openings, adding an extra step to the process.
- Pest control is needed.
- Security in tight urban settings surrounded by hungry people can be an issue. Some micro-farmers use crates and move them indoors at night.

YOU
- Enthusiastic reports of sack gardening's successes are anecdotal. Studies are needed to document best practices and quantify impact on food security and local economies.

EDU
- Learn how to implement a gardening project by enrolling in the Center for Sustainable Development's online course "Food Security, Nutrition & Home Gardens."

YOU EDU
- Instructions for creating sack gardens are posted at Appropedia.org.

Improved Crop Storage Bags

Airtight triple bagging provides safe long-term storage for cowpeas and other crops.

tel

postharvest.org • @Postharvest.org

One Acre Fund farmer Agnetta Musonye shows her maize harvest, stored in PICS bags, Shimanyiro, Kenya. © Kelvin Owino

Subsistence farmers toil daily, under harsh conditions, to raise crops. Yet an estimated 30 percent of yields are lost post-harvest, suppressing revenue and exacerbating food insecurity.

Cowpeas, called black-eyed peas or *niébé*, are important in central and West Africa, and are grown primarily by women.

Cowpeas' desirability derives from their:

- ✓ Drought resistance, shade tolerance, and ease of growth
- ✓ Quick growth as a ground cover, helping to prevent soil erosion
- ✓ Nitrogen-fixing properties, which improves the soil
- ✓ High-protein content, which makes them extremely nutritious
- ✓ Stems, leaves, and peas being fully consumable by humans, as well as by livestock

Cowpeas are highly vulnerable to infestation, unfortunately. After several weeks in granary storage, bruchids, also known as cowpea weevils, have laid eggs in the cowpeas. Within a few months, the eggs hatch, rendering the cowpeas worthless. To avoid this, farmers sell their cowpea crops at harvest time—just when supplies are highest and prices are the lowest.

Purdue University, in partnership with African scientists and farmers, searched for a cowpea storage solution. In 2007, they introduced affordable triple-decker **Purdue Improved Cowpea Storage (PICS)**, bag systems. Airtight bags hermetically seal the cowpeas, killing the insects without using pesticides or other chemicals, and do not affect the quality of the crop. PICS bags:

- ✓ Consist of two heavy inner bags, one fitting inside the other, that fits inside a third woven, tear-resistant outer bag
- ✓ Last several years, amortizing their $2–$3 cost over many harvests
- ✓ Are manufactured in Nigeria and sold directly to farmers by agro-vendors
- ✓ Protect contents for up to a year, maximizing farmers' revenue by allowing them to sell when supplies are lowest and prices rise

The PICS initiative covered ten West and central African countries. Millions of farmers, a large percentage women, have improved their bottom lines by adopting this low tech/high impact innovation.

> "The bags were so successful that one farmer's organization, the Sabati Women's Association, in Zantiebougou, Mali, decided to buy them directly from the supplier. 'The women saw how well they worked and wanted to buy the bags themselves,' says P4P [Purchase for Progress] regional coordinator Isabelle Mballa."
>
> —*Eliza Warren-Shriner, World Food Programme*

Users demonstrate their Purdue Improved Crop Storage Bags, Walewale, Ghana. © Dieudonne Baributsa

The Bill and Melinda Gates Foundation is funding further studies at Purdue, testing PICS bags for other crops. They have been renamed **Purdue Improved Crop** [no longer "cowpea"] **Storage** bags.

Guaranteeing bag supply to remote farming regions is not a slam dunk. Vendors' profit margins on the bags are small, and since the bags can be reused many times, demand fluctuates.

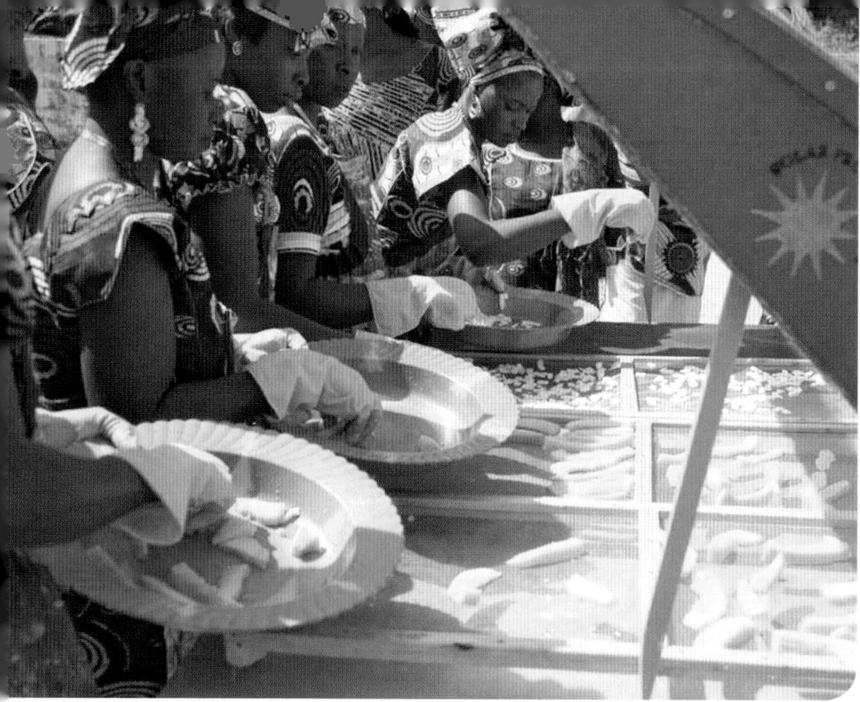
The Gambia Women's Initiative dries mango and coconut slices in their solar dehydrator in Samita, The Gambia. © Lonnie J. Angstadt/Power Up Gambia

Solar Dehydrators and Nut Shellers

Improved crop-processing techniques preserve yields and save farmers' time. Solar dehydrators and nut shellers are affordable when shared by women's cooperatives.

D·I·Y H L&G tel

powerupgambia.org • @PowerUpGambia • TheFullBellyProject.org • @FullBellyProj

68 Some crops mature at staggered intervals, but many ripen all at the same time, requiring concentrated labor to harvest them, transport them to the marketplace, and/or store them. This is the point when high percentages of crops are lost. Mangoes, for example, all come to market at roughly the same time, lowering prices. In The Gambia, up to half the mango crop is wasted due to the lack of refrigeration or safe storage. Juice production requires an industrial supply chain only just being developed.

Air-drying food, an ancient technique, is hindered by:

✗ Insects or animals feeding on the produce
✗ Rain or high humidity
✗ Dirt and contamination

Low-tech solar dehydrators provide an effective method to preserve produce, offering many benefits:

✓ Dried food decreases in volume and weight—up to 95 percent of water evaporates—easing storage and transport.
✓ Removing moisture from food prevents bacteria, yeast, and mold growth while retaining nutrients.
✓ If markets can be accessed, dried produce sales are viable microenterprises.
✓ Dried mango improves local diets.

The Gambia Women's Initiative has introduced solar dehydrators to dry fruit both for local consumption and off-season sales. Calorie-dense dried mangos, rich in vitamin A, combat vitamin A deficiency (**#7**).

Solar dehydrators run strictly on the sun and are low-tech construction projects utilizing local materials. The flat, large platform versions are generally owned by women's collectives. Solar heating speeds up the evaporation of the fruits' and vegetables' water content, with natural ventilation drawing out moist air.

- Clear glass or plastic covers protect food while letting sunlight through and trapping the heat.
- Black interior surfaces absorb the sun's rays.
- Foods, whole or sliced, are laid out on inner mesh trays, allowing 360 ° air circulation.
- Most fruits and vegetables require 1–3 days to dry.

Three Indian grad students, Vaibhav Tidke, Shital Somani, and Aditya Kulkarni, won the 2013 Dell Social Innovation Challenge grand prize for their sleek **solar conduction dryer**. At $57, it will quickly pay for itself through sales of dried produce and fish otherwise wasted.

The **Universal Nut Sheller** was originally developed by **The Full Belly Project** for a Mali women's cooperative. Human-powered, it shells peanuts thirty times faster than doing the job by hand. Machine shelling protects women's hands from the ill effects of the aflatoxins that peanut shells can harbor.

Shelled peanuts fetch a much higher price, so it is value-added tech. The peanut shells can be used in eco-briquette (**#32**) production, saving wood-gathering time. The UNS also works with other cash crops like shea, jatropha, and coffee.

See also: **Wheeled Seeders and Paddy Threshers (#61)**

Agroforestry: Integrating Trees into Subsistence Farming

Trees provide powerful ecological and economic benefits.

TreesForTheFuture.org • @TreesFTF

Young girls haul wood in Chuiquel, east of Lake Atitlan, Guatemala. © Alejandro Guarcha/CYE Global Library

Trees' ecological and economic benefits—both local and global—include providing:

- ✔ Erosion prevention, supporting soil stability and water retention through deep rooting
- ✔ Wildlife habitat, nurturing birds, bees, butterflies, and other beneficial fauna
- ✔ Food and natural resources for domestic and commercial use
- ✔ Protective shade and windbreaks
- ✔ Carbon absorption, mitigating global warming

Trees are under global assault, felled for cooking fuel or other consumption, and clear-cut for industrial farming. Denuded, depleted lands imperil subsistence farmers' livelihoods.

Agroforestry, integrating trees into subsistence agriculture to restore degraded land, helps low-income farmers improve and diversify their yields, raise household income, and improve health. Many NGOs and intergovernmental organizations focus on identifying the most productive, suitable trees for different climates and promote their distribution and cultivation.

Trees for the Future deploys a variety of first-stage trees to communities who request their help.

> "We initially plant multipurpose, fast-growing trees that lower daytime soil temperatures, provide partial shade, and add humus and nutrients to the soil. This creates a microclimate that allows dormant seeds and native trees to germinate and reestablish themselves, which helps to restore much of the former diversity."
>
> —*Trees for the Future*

A farmer with moringa seedlings (and a water filter), Blufields, Nicaragua. © Peter Fosdike/ Moringa Harvest.co.uk

To be maximally effective at this, fast-growing, multi-purpose trees must:

- ✔ Survive in full sunlight (twelve hours)
- ✔ Have strong taproots
- ✔ Be coppiceable (regenerating vigorously when cut)
- ✔ Grow leaves useful for animal forage, organic fertilizer, or other purposes
- ✔ Fix nitrogen

While many trees fit those criteria, the first among equals is the ***Moringa oleifera***. Moringa trees are nutritional powerhouses, highly recommended by professionals working to prevent malnutrition and micronutrient deficiencies (**#7**). Packed with ninety-plus vitamins, minerals, vital proteins, antioxidants, and omega oils:

- ✔ Moringa is drought-tolerant, growing in arid tropical and subtropical regions throughout Asia, Africa, and Latin America—all high-malnutrition regions.
- ✔ Moringa grows quickly, nine to fifteen feet a year.
- ✔ Moringa produces oil for cooking and high-value commercial use; oilseeds are rich in antioxidants, beneficial for hair and skin.
- ✔ Moringa leaves are used in making cosmetics and personal care products.
- ✔ Moringa leaves can be consumed raw, powdered, or cooked, providing:
    - 7 times more vitamin C than oranges
    - 3 times more iron than spinach
    - 3 times more potassium than bananas
    - 4 times more vitamin A than carrots
    - 4 times more calcium than milk
    - All the essential amino acids needed to build strong, healthy bodies

The Philippine **Moringaling.net** is devoted to all things moringa, hosting the international

The drought-resistant Sahel apple trees provide a vitamin-rich fruit along with the benefits of afforestation. © ICRISAT

Preparing moringa leaves before cooking in Konso, Ethiopia. © Trees for the Future

Moringa Congress to share research on cultivation, commercial supply chain development, and promoting moringa's wide range of assets.

Kwami Williams and Emily Cunningham co-founded **MoringaConnect.com** after traveling to Ghana with MIT's D-Lab. Their social enterprise is developing a moringa oil processing industry in Ghana, connecting smallholder farmers to world markets. Moringa seeds can be harvested just one year after planting.

Experiments utilizing moringa seeds for water treatment (**#42**) have been promising.

Leucaena is another top performing, fast-growing multi-purpose tree:

- ✔ Self-pollinating, easily reproducing
- ✔ High-nitrogen content fallen leaves are excellent fertilizer
- ✔ Effective as a live fence due to its dense leaf-cover
- ✔ Regenerates when cut for fuelwood and timber
- ✔ Leaves provide animal forage

Calliandra trees are also widely used in agroforestry, due to their many assets:

- ✔ Excellent firewood, resprouting after cutting
- ✔ Leaves are high in nitrogen, fertilizing when they fall
- ✔ Leaves can be used as livestock fodder—goats and cattle like it
- ✔ They are especially effective as a pioneer species to reclaim over-used lands, thriving on slopes and marginal soils
- ✔ They produce nectar-rich flowers, facilitating year-round honey production (**#70**)

The gliricidia tree, planted in between maize rows (**#60**), greatly increase yields. It is useful for live fencing, fodder, coffee shade, and firewood. It fertilizes when it drops its nitrogen-rich leaves.

One Acre Fund distributes grevillea tree seeds to each smallholder farmer in its program. Quick-growing trees requiring little water, they rejuvenate the soil and can be sold as fuel.

> Kenya's **Greenbelt Movement,** started by the late Professor Wangari Maathai, is a world-famous tree planting initiative. Under her leadership, millions of trees were planted in Kenya. Maathai was awarded the 2004 Nobel Peace Prize for her commitment to habitat restoration through reforestation.

Growing trees for fuel wood positively impacts girls. Trees that regenerate after cutting are a renewable source of on-site firewood. Girls' wood-fetching duties are therefore accomplished in far less time, improving their school attendance. This benefits the girls, their families, and their entire communities. Pairing sustainable firewood cultivation with **improved cookstoves** (**#50**) lowers wood consumption, further lightening girls' loads.

See also: **Intercropping** (**#60**)

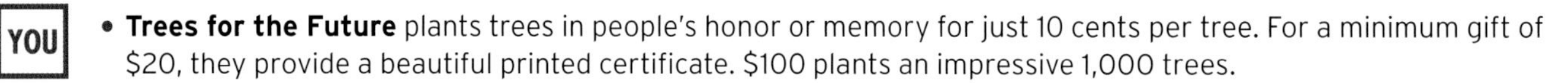

YOU

- **Trees for the Future** plants trees in people's honor or memory for just 10 cents per tree. For a minimum gift of $20, they provide a beautiful printed certificate. $100 plants an impressive 1,000 trees.
- **COTAP.org**, Carbon Offsets to Alleviate Poverty, offers tree planting to offset your wedding's carbon footprint.

EDU

- **Trees for the Future** offers a free online agroforesty mini-course. Those who pass the exam are eligible to sponsor projects and receive technical assistance from TFTF.

Beekeeping

Beekeeping is relatively inexpensive to set up, yields honey and byproducts, and offers local and global benefit.

UTMT.in • @UTMT • HoneyCareAfrica.com • @HoneyCareAfrica

Bee trainers Bharati, Usha, and Malutai in Kisrul, a village in Maharashtra, India. © Under the Mango Tree

In addition to providing cash crops, honey and beeswax, bees can improve small-plot farms' productivity through increased pollination. In warm climates, bees produce two or three honey harvests a year.

Well-suited for female subsistence farmers, beekeeping:

- ✔ Requires little land. Bee boxes are compact and can be mounted aboveground.
- ✔ Requires only modest amounts of training and technical assistance.
- ✔ Only requires a one-time expense: the cost of modern, higher-producing Langstroth Hives (rather than traditional versions), bees, and a few tools. Bees do not require inputs or food, just access to water. The first harvest pays off the loan, and the hives continue producing honey.
- ✔ Improves crop yields 15–30 percent through greater pollination.
- ✔ Diversifies income streams without time-intensive labor.
- ✔ Produces honey, which does not spoil, so it is low-risk.

"During our first annual general meeting . . . the women said this venture is a poor rural woman's savior."
—*Co-founder Rose Akaki, Maruzi Bee Keepers' Association, Uganda*

Honey is flavored by the bees' floral nectar source. Monofloral honeys like orange blossom, cardamom, litchi, and sweet clover fetch high prices. Wildflower honey, from mixed floral sources, is also a valuable commodity. Beeswax is prized for candles and as an ingredient in salves and creams.

Under the Mango Tree, an Indian social enterprise founded by Vijaya Pastala, capitalizes on the high-end niche honey markets by developing a "plow to plate" supply chain. Her business connects local subsistence farmers with larger markets, producing and marketing their gourmet Under the Mango Tree–branded honey.

Under the Mango Tree runs an urban beekeeping program in Mumbai, offering locals apiculture training.

Honey Care Africa, a Nairobi-based social business, likewise links smallholder farmers to larger East African markets. Honey Care Africa builds its own hives for Business-in-a-Beehive participants, funded through their microcredit program. The company provides ongoing training and education; some of the teachers are their own alumni.

Fair trade (**#93**) honey prices are set in advance, allowing producers to pay off their loans with the sale of their honey harvest. Honey Care Africa partners with Kiva.org, a personal microfinance lending platform.

Massive global bee die-offs have set off environmental alarms and much scientific research is under way to determine and—hopefully—correct the underlying causes. Bees are free-range, so beekeeping reduces the local economic incentives for deforestation, encouraging subsistence farmers to focus on the bigger ecological picture to ensure plentiful bee foraging sources.

Challenges:

- Bee populations can plummet or die off entirely, as is happening worldwide.
- Bee sting fears can scare off prospective beekeepers.
- Accessing markets beyond their immediate rural area is always challenging for female farmers.

See also: **Microloans** (**#88**), **Fair Trade** (**#93**)

- Intern in Mumbai with **Under The Mango Tree**.
- Plant locally appropriate wildflowers to provide forage for bees.

© Gardens For Health, Rwanda

Chickens and Eggs

Poultry farming provides eggs and meat for sale, as well as for family consumption.

GardensforHealth.org • @Gardens4Health • KayeemaFoundation

71 Chickens, ever-popular entry-level livestock, offer potentially high returns on investment and require just basic skills and a chicken coop. Chickens eat food scraps and foraged bugs plus supplemental feed provided by the farmer. Poultry rearing, traditionally a women's enterprise, has many merits:

- ✔ Chickens require little space.
- ✔ Chickens are relatively low-maintenance.
- ✔ Well-fed hens lay up to 200 eggs a year, adding protein to the family's diet.
- ✔ Fertilized eggs hatch in twenty-one days, so flocks increase quickly.
- ✔ Chicks expand farmers' flocks or are sold.
- ✔ Surplus chickens are a flexible asset that can be controlled by women, used as food or cashed out.
- ✔ Chicken manure added to compost enhances vegetable growth.

Gardens for Health, a Rwandan-based NGO, serves many female-headed low-income households. They offer participants training and inputs for vegetable gardening and chicken farming.

Francine, a recipient of two chickens from Gardens for Health, tells her story:

> *"One of the chickens . . . produced seven chicks; I kept them in house until grown without being eaten by any animal. For the second time, this chicken had eight baby chickens and again, I raised them in house . . . For now I have forty-two chickens . . . Every time someone comes to my house and asks me for a chicken either for rearing or eating I always give because I have got them by generosity . . . I do not only get eggs and meat from chickens, I also get chicken manure for the vegetable garden I grow . . . A chicken is a domestic animal that does not require a lot to grow. When you care about it, feed it, give it medicine when in ailment period, and protect it from animals which eat it, it produces much more than you imagine. It leads you and your children to better living conditions."*

> "You may rear chickens to feed your family, sell them or eggs on markets for money or even buy another type of livestock if you want to. And this can all be done from only two chickens . . ."
>
> —*Gardens for Health International Blog*

Kuroiler chickens, bred in India, lay more eggs and grow larger than indigenous breeds. They can thrive on a diet of agricultural and household waste, requiring no supplemental feedings.

© Gardens for Health, Rwanda

Challenges:

- Hawks, foxes, dogs, and other predators—including thieves—can kill chickens and steal eggs
- Veterinary services that provide vaccinations and treatment for diseases may be difficult to access. The larger the flock, the higher the disease risk. Utilizing modern methods increases the likelihood of successful poultry farming. The Kayeema Foundation trains para-vets, focusing on effective rural poultry vaccination programs.

YOU

- Contact **Gardens for Health** about volunteering in Rwanda.

Goats

Goats, traditionally raised by women, provide milk to improve family diet and income from selling milk and offspring.

HealthinHarmony.org • @Health1nHarmony

Francine, a beneficiary of the Good Gifts initiative goats-for-peace in Ntarama, Eastern Rwanda. © Andrew Sutton/ Survivors Fund/Goats from Good Gifts

72

Goats, a central fixture of subsistence farming, thrive on free-range forage. Goats are traditionally managed by women, providing them with a source of income. Goats can produce two or even three offspring a year, flexible assets sold to cover food, school fees, farming expenses, health emergencies, housing upgrades, or even dowries.

Heifer Fund International and many other NGOs sponsor goat programs. Goats are photogenic and relatively inexpensive. People are encouraged to underwrite goats for impoverished families in lieu of material gifts.

Heifer Fund's most famous goat recipient, Beatrice Biira, was raised in a Ugandan family that received a Heifer Fund goat. It did indeed help their family earn enough to send the children to school. Beatrice became a star student, receiving scholarships to attend secondary school and college in the United States. She now works for Heifer Fund. Her story is retold in an illustrated children's book, *Beatrice's Goat.*

- ✔ Traditionally raised goats are low-input, foraging for their food.
- ✔ Goats flourish in harsh climates and survive droughts.
- ✔ Dairy goats provide milk, yogurt, and cheese for families to consume or sell.
- ✔ Surplus offspring can be sold, fetching highest prices at holiday times.
- ✔ Goat manure promotes vegetable growth and is sold as organic fertilizer.

Vikram Akula explained "Goat Economics" to Bill Gates and Warren Buffet—why borrowing even at a relatively high interest to buy goats makes sense:

> *"... a landless agricultural worker might use a 2,000 rupee loan (about $40) to buy a goat. She continues with her daily work and takes the goat along with her to the fields. The goat eats grass and virtually anything else, so there is no investment from her end. A goat gives birth to one or two kids a year and the value of the offspring is about 50 percent of the mother, or about 1,000 rupees. Even if a borrower took a 28 percent loan, she makes a return of about 70 percent on invested capital."*

Health In Harmony, based in Borneo, supports conservation-friendly poverty-alleviation programs, decreasing pressure on Gunung Palung National Park. They give goats to widows, among the most marginalized and impoverished community members. The widows then sell offspring and organic goat manure, gaining livelihood and community respect.

Challenges:

- Veterinary services for the goats need to be accessed and paid for.
- Untethered goats graze indiscriminately; without supervision they can eat valuable plants and young trees, denuding landscapes and causing conflicts between neighbors.
- Paradoxically, children tasked with goat-tending may miss school.

See also: **Microloans (#88)**

YOU

- **VetsWithoutBorders.ca**, based in Canada, runs trips for veterinarians to share their expertise, working with farmers and their livestock in Uganda and elsewhere. VWB also runs international programs for veterinary students.

Adivasi women fish-cage farming in Bangladesh. © Mehadi/WorldFish

Fish Farming

Fish farming provides high-protein food for the family and surplus fish to sell.

D·I·Y H L&G

WorldFishCenter.org • @WorldFishCenter • PracticalAction.org • @PracticalAction

73 Creating microfisheries opens opportunities for women living near rivers, lakes, deltas, canals, and coastal waters. Overfishing is depleting ocean stocks; prices are rising. Aquaculture provides affordable, tasty, high-protein fish.

Start-up costs are modest. Fish cages or ponds are adjacent to home, allowing women to integrate fish farming with their other work. Often women's groups share fish-farming duties.

Fishponds are dug and stocked with tilapia, catfish, carp, and other types of fish fingerlings. Mola, small micronutrient-rich indigenous fish, can be raised along with other fish and harvested frequently for household meals. Fish, together with vegetables grown along the pond's banks, diversify familes' rice-based diets, combating malnutrition and micronutrient deficiencies (**#7**).

> "The extra money I gained through fish farming helped me to buy food for my family. Some of the money I used to reinvest in the fish and vegetable culture, and I placed some in the bank to prepare for any future crisis."
> —*Sufia Begum, Aquaculture for Income and Nutrition participant*

Practical Action has initiated successful fish-farming programs in Bangladesh:

- A few young fish are put into each "hapa" [cage], which acts as their home, floating just below the surface of the pond. All they need is a little food—oil cake, duckweed, kitchen waste and snails—and in just a few months they grow to full size. Then they produce more fish, and more, and more.
- Cages are made using a few cheap materials. Bamboo poles form an outer frame that is covered in netting; a "nursery" section is created inside for the younger, delicate fish; floats are added at the corners.
- One cubic meter, the cage holds up to 300 fish at a time . . . for two growing seasons each year.

Cages cost under $10 and last a few years. Once families learn the technique, they can continue fish farming independently, and neighboring people have adopted it. Fish can be dried in solar dehydrators, for future consumption or sales (**#68**). Seaweed is also grown in cages and can filter marine water.

Katrin Puetz of B-Energy, the inventor of BioGas bags, reports that slurry, the byproduct of biodigesters, can be used as fish food.

A fish farmer using a lift net in her fish pond in Rangpur, Bangladesh. © Holly Holmes/WorldFish

Challenges:

- Industrial fish farming can be highly polluting. Microfisheries are more sustainable, but care must be taken to avoid ecological damage.
- Fish farmed in public waters are vulnerable to theft; some communities share guard duty.
- Polluted water runoff or aquatic weed growth can harm fish.

photo (facing page): Baifikrom Primary School students plant a cassia tree in Mankessim, Ghana. © Earth Child Institute/Ghana ▶

YOU • **SaelaoProject.com**, a sustainable community in Nathon, Laos, includes fish farming and welcomes volunteers.

Sector 7 – An Overview

CONSTRUCTION

The world's rural poor generally own their humble houses. In the sprawling favelas, shantytowns, slums, and informal settlements – technically known as peri-urban because they are adjacent to formal city limits – slum residents rarely own their ramshackle homes. Some are squatters, others are tenants. Few have deeds to homes, a disincentive to improving them. The houses may be destroyed and the residents evicted at any time.

What the rural and urban homes of the poor have in common is a lack of basic infrastructure. Often not much more than makeshift walls with flimsy roofing, they rarely have:

- Running water for consumption or plumbing
- Toilets; latrines are a major upgrade
- Electricity, unless residents own portable solar lanterns or have installed a solar panel
- On-demand cooking fuel like piped gas
- Ventilation, cooling, or heating; indoor smoke from cooking and kerosene lamps is the norm

Inferior housing and poverty are related. This book includes many techniques to improve quality of life. When a family's income rises, they typically move or improve their living quarters. A new house is well beyond $100, but many small upgrades are affordable. Dr. Pilar Mateo (**#77**) emphasizes the poverty trap of poor housing that causes poor health; without resources and access to education, it is harder for people to protect themselves, exposing them to illness and damage, both of which block progress out of poverty.

This small sector is comprised of innovative techniques, three based on creative reuse of plastics. The world is awash in plastic trash. Upcycling it into construction materials and built-in daytime lighting is an eco-smart win-win.

74. BOTTLE BRICKS/ECOBRICKS

Plastic bottles filled with trash or dirt make strong, durable bricks that build sturdy constructions.

75. IMPROVED ROOFING

Traditional low-cost roofing options leave much to be desired. Upcycling plastic bottles into thatch is an eco-friendly, affordable, and effective roofing innovation.

76. PLASTIC BOTTLE SOLAR LIGHTS

Plastic soda bottles filled with water and inserted into the roof refract light equivalent to a 55-watt bulb.

77. INESFLY: INSECTICIDE-INFUSED PAINT

Dr. Pilar Mateo invented insecticide-infused paint, protecting people from deadly Chagas disease (spread by vinchuca beetles); it shows promise fighting other insect-borne diseases, too.

A shopkeeper with a new solar plastic bottle light, Manila, Philippines. © Joey DeLeon

Building a bottle school in San Martín Jilotepeque, Chimaltenango, Guatemala. © Hug It Forward

Bottle Bricks/ Ecobricks

Plastic bottles filled with trash or dirt make strong, durable bricks to build sturdy constructions.

D·I·Y L&G S2N tel

HugItForward.org • @HugItForward • CasasConBotellas.com • @PuraVida.org • Ecobricks.org

Trashed plastic bottles and litter cover the planet worldwide. Discarded bottles filled with garbage can be **upcycled** as **ecobricks**, cleaning up communities and providing valuable building materials. Due to the absence of waste management infrastructures and the scarcity and high cost of construction materials—especially when shipped to remote areas—bottle bricking has taken off around the world.

German architect Andreas Froese pioneered a horizontal dirt-filled bottle brick technique in Honduras in 2005. Around the same time, Susanne Heisse—founder of **Pura Vida**, a movement for alternative trash management—realized her ecobrick creations could be used to rebuild after hurricanes destroyed much of their home base near Lake Atitlan, Guatemala. Her method utilizes bottles filled with inorganic trash, vertically stacked and held in place by a wood framework and chicken wire.

Bottle brick construction is community-based, creating shared structures. "Bottle schools" have become popular in Guatemala, inspired by Peace Corps volunteers who picked up the idea from Pura Vida. Peace Corps Volunteer Laura Kutner led a community construction process that made 6,000 bottle bricks to build a local school. Bottle school projects are now included in Peace Corps training.

- ✔ Using ecobricks and donated labor, bottle schools cost only a fraction of what municipalities budget for school buildings.
- ✔ African bottle brick construction includes latrines, biodigesters, homes, and even a whole Nigerian housing project.
- ✔ The thermal mass of the lateral technique's thick walls retains cool air in hot seasons and maintains warmth in colder stretches.
- ✔ Bottle bricks' carbon footprint is zero, unlike conventional bricks fired at high temperatures.

Eliza Moreno blogged from the St. Monica's Tailoring School in Gula, Uganda, where she spearheaded a bottle brick home-building project:

> *"The construction process of building with bottles is work intensive. This means many can be involved in the process, creating opportunities for employment and community involvement, from collecting to filling to building. In our case, students from the area were invited to fill bottles in exchange for scholastic materials such as notebooks and pencils . . ."*

In the lateral method, bricks are set in cement, adobe, or **cob mortar**, a mixture of clay, sand, straw, and water. In both vertical and lateral methods, bricks are plastered over, creating smooth walls with no hint of the bottles within.

It is important to use only non-biodegradable trash for filling like wrappers or other plastic waste. The bottles are collected before construction; organic fill can mold during storage.

Exterior details add visual interest:

- ✔ Sometimes builders leave the curved bases of the bottles visible, creating attractive patterns and shapes.
- ✔ Partially exposed tinted bottles, or clear bottles filled with colored water, provide beautiful stained-glass effects, especially if light shows through them.
- ✔ Bottle caps can be pressed into the wall's cover coat to create mosaic-style decorative motifs.

Bottle bricking is well-suited for incremental construction, whereby people expand their homes or sheds piecemeal as time and resources permit. Bottle bricking is a low-tech, low-barrier, do-it-yourself construction technique that women can easily learn and apply.

Jane Liwan, an elder in the small Philippine town of Besau, has rebuilt all her home's walls with ecobricks, using colored bottles so artistically that people throughout the province come to see her home.

Lateral bottle bricking headed up by teacher Rebecca Bacala in Mt. Province, Philippines. © Rebecca M. Bacala/Ecobricks.org

Students attending school adjacent to western Uganda's Kibale National Park prepare bottle bricks to build a school biodigester. © Kate Wrangham-Briggs

Ingrid Vaca Diez balances on the scaffolding at one of her bottle brick home-building projects. © Ingrid Vaca Diez

Bolivian Ingrid Vaca Diez, a lawyer with a passion for helping the poor, is famous for her bottle brick home-building projects. When her husband complained that the mountain of discarded bottles on their patio was enough to build a house, she did just that. After finishing up her first bottle-brick home, she kept on going. She has headed up ten bottle-brick teams, building houses with low-income families.

Hug it Forward builds bottle schools in Guatemala, inspired by Peace Corps volunteers' projects. They use the vertical Pura Vida technique, organizing teams of volunteers to collaborate with locals to build village schools.

Hug It Forward stresses the environmental benefits of building with trash-filled bottle bricks. Because rural Guatemala lacks waste disposal infrastructures, trash is often burned, releasing toxic fumes, or it is dumped in waterways, where it contaminates beaches, lakes, and rivers, and damages fish and wildlife.

> "After two years of unfolding in Mt. Province, Philippines, ecobricking has become a long-term community habit. Dump sites are used less or have completely shut down! Plastic burning and tossing have been dramatically reduced . . . Some villages have so little plastic, they borrow from their city friends so ecobrick parks can be finished."
>
> —*Russell Maier, Guide to Eco-Bricking*

YOU

- Join **Hug It Forward** on a weeklong trip to Guatemala, helping build a bottle school. They welcome both individuals and families.
- **Pura Vida** hosts work weeks; volunteers cover their own expenses.

EDU

- **Ecobricks.org** has a downloadable *Guide to Eco-Bricking*. Lead author Russell Maier heads up community projects constructing beautiful bottle brick eco-parks where he teaches in the Philippines.

EDU

- **The Peace on Earthbench Movement (POEM)** promotes public ecobrick bench construction projects around the world. Check out their ecobricking guide and ideas: **www.earthbench.org**
- Build a keyhole garden (**#65**) with bottle brick walls, combining the eco-benefits of both innovations.

Vananh Le demonstrates plastic thatch roofing, piloted in Ecuador.

Improved Roofing

Traditional low-cost roofing options leave much to be desired. Upcycling plastic bottles into thatch is an eco-friendly, affordable, and effective roofing innovation.

L&G tel

ReuseEverything.org

Putting a roof over your head is a basic human goal, even if it's plastic tarp or cardboard. Traditionally, women have gathered grasses for thatched roofs. While picturesque, thatched roofs have significant downsides:

- ✗ Thatching is time-consuming and needs continual maintenance; it can leak and collapse.
- ✗ Repairing and replacing thatch is costly. Grasses for thatch are becoming scarcer.
- ✗ Thatch provides habitat for unwelcome creatures, such as disease-spreading insects, small animals, and worms.
- ✗ Dry thatch is flammable.

Consequently, when families can afford to do so, they upgrade their roof. Corrugated metal sheeting is the most commonly available material, virtually a defining feature of worldwide low-income housing. Metal roofs are valued for being:

- ✔ A solid investment, requiring little maintenance
- ✔ Watertight
- ✔ Gutter-friendly, facilitating rainwater harvesting, and handily providing natural racks for SODIS, solar disinfection using water bottles placed in direct sunlight (**#39**)

Unfortunately, corrugated roofs have significant drawbacks, in addition to being a large expense for impoverished people:

- ✗ They are heat-trapping; since the world's poorest live primarily in hot climates with no electricity to run air conditioning or fans, this causes major discomfort.
- ✗ They are noisy when it rains, drowning out all conversation, though there are techniques to dampen the din.
- ✗ They are not aesthetically pleasing.

GIveDirectly.org disburses cash to poor households without stipulations on how they can spend the money. GiveDirectly's working hypothesis is that poor people's priorities are rational and instructive. Evidence shows recipients use their money to improve their financial stability. One of the first things many do is upgrade from thatch to metal roofing, saving time and money.

Dr. David Saiia co-founded the **Reuse Everything Institute** to combine sustainable reuse of waste materials while providing economic opportunity. He has developed plastic thatch, an innovative roofing material. This roofing material combines the advantages of thatch:

- ✔ Sound muffling
- ✔ Aesthetic appeal
- ✔ Air circulation

With the benefits of plastic:

- ✔ Cheap—or free—and readily available
- ✔ Long-lasting—bad when it's litter or landfill, but good for roofing
- ✔ Translucent, so home interiors enjoy natural lighting; plastic thatch is clear plastic or variegated, using a mix of different colored bottles
- ✔ Waterproof

Saiia, along with co-founder Vananh Le and the Carnegie Mellon **Engineers Without Borders** chapter, is refining a manually operated machine to create thatch strips from discarded soda bottles.

Local thatch-producing businesses—upcycling discarded plastic beverage bottles and creating jobs while improving low-income housing—will be **triple bottom line** endeavors: benefiting people, friendly to the planet, and generating profits.

> **Plastic thatch collects enough soil over time to become a naturally occurring green roof, extending the life of the plastic. In Ecuador, thatch roofs have sprouted orchids.**

YOU

- Reuse Everything Institute seeks **impact investment**.

Plastic Bottle Solar Lights

Plastic soda bottles filled with water and inserted into the roof refract light equivalent to a 55-watt bulb.

tel

ALiterOfLight.org • @ALiterOfLight

A shopkeeper with a new solar plastic bottle light, Manila, Philippines.
© Joey DeLeon

Brazilian mechanic Alfredo Moser was frustrated when power went out at his workplace frequently. He started playing around with natural lighting solutions and invented plastic bottle solar lights. A sealed bottle full of water is installed in an overhead roof, a portion above the roof and the balance below. Refracting the light shining through it, it illuminates without electricity.

These only work when there's sunlight, of course, but many slum shacks have no windows, making daytime light a valuable, nearly free upgrade. This source of light requires just a few steps:

- A clean bottle is filled with water and two tablespoons of bleach (the bleach is added to prevent bacteria buildup).
- A circle is cut out of a square sheet of metal and a matching hole is cut in the roof.
- The capped bottle is inserted into the sheet metal base and pushed through the hole.
- A rooftop, watertight seal is created around the bottle and base.

Lighties, invented by South African Michael Sutton, are just launching. They add an interior tube housing a bulb and cylindrical solar charging panel to a water bottle light bulb. They provide light at night, too.

Solar bottle lights are a low-tech, high-impact innovation. They have traveled around the world within just a few years with the help of **social entrepreneur** Illac Diaz, head of Philippine-based **Liter Of Light**, and via social media.

Humanitarian technology transfer, the spreading of tools and practices appropriate for low-resource areas, is generally envisioned as flowing from high-resource areas to the developing world. The industrialized world, after all, has capital, resources, and a vast network of intergovernmental organizations, NGOs, universities, and philanthropies.

Several significant innovations have originated in the developing world and transferred to the industrialized world, though—what some call reverse tech transfer. It makes sense that global southerners are innovative problem-solvers: they live with constrained resources. The Internet has helped spread their unique solutions. For example, kangaroo care for premies (**#2**), developed in Colombia, has become a standard component of affluent world premie protocol.

Plastic bottle lights represent a new stage of do-it-yourself tech transfer: from one Global South region direct to another, transmitted via social media rather than through any specific effort or campaign. Once people learn of an effective technique, directions for implementation are readily available. Bottle bricks (**#74**) originated in Guatemala and have traveled to the Philippines. Visual directions bridge language barriers. Two rich sources (among many) are:

- ✔ **Instructables.com**—a spin-off of the MIT Media Lab, is a web-based documentation site where more than 100,000 projects have been shared, including photographic instructions.
- ✔ **Appropedia.org**—is a collaborative effort to share information about sustainability, **appropriate technology**, and poverty reduction.

YOU EDU

- Directions are posted at ALiterofLight.org.

Dr. Pilar Mateo is the inventor of Inesfly, an insecticide-infused paint.
© Inesfly-Bolivia

Inesfly: Insecticide-Infused Paint

Dr. Pilar Mateo invented insecticide-infused paint, protecting people from deadly Chagas disease (spread by vinchuca beetles); it shows promise fighting other insect-borne diseases, too.

PilarMateo.com • @PilarMateoh • Inesfly.com • @inesfly • MOMIM.org • @momim_momim

77 Spanish paint chemist Dr. Pilar Mateo has become a global humanitarian through her invention of an insecticide-infused paint called **Inesfly**. When applied on a house's exterior, it helps to protect rural, impoverished Bolivians from nighttime vinchuca bites. These blood-sucking, vampire-like beetles live in crevices of the mud walls of indigenous peoples' homes. Vinchucas infected with the parasite that causes Chagas disease transmit it to humans.

When bitten by an infected vinchuca, victims initially experience a bout of flu-like symptoms. Then Chagas disease lies dormant, eventually erupting and damaging the digestive system and heart, sometimes fatally. In Latin America, the epicenter of the vinchucas' territory, an estimated 25 million people are infected.

Mateo's Valencia-based family owned a paint factory. She earned her doctorate in paint chemistry and joined the family business. When a local hospital was closing down due to a cockroach invasion, she had an idea. Her innovation—insecticide-infused paint, toxic for roaches but not to humans—was a success.

Word spread and a Bolivian activist helping impoverished indigenous Bolivians combat chronic vinchuca infestation invited Mateo to visit and try her technique there. She accepted and her product has worked as intended. Inesfly effectively "vaccinates" houses, rather than people. Deploying Inesfly reduced infestation rates from as high as 90 percent to nearly zero.

Mateo's micro-encapsulated paint slowly releases insecticide.

- ✔ Because just small amounts of insecticide are gradually released, the paint is less toxic than fumigation. (Inesfly uses a mix of WHO-approved insecticides).
- ✔ Treatments last two years; spraying needs to be done twice yearly.
- ✔ Comprised of both insecticides that kill mature insects and insect growth regulators, which kill eggs and young insects, it reduces overall insect population.
- ✔ It is proving effective in fighting malaria and dengue fever, diseases that are transmitted by mosquitoes.
- ✔ It is approved for use in more than fifteen countries.

Inesfly has opened a second factory, in Accra, Ghana, producing paint that protects against malaria, central to African public health. Manufacturing nearer to purchasers lowers labor and transport costs and builds local capacity.

Challenges:

- Inesfly does not work on thatch, common among the poor.
- It is ineffective against pesticide-resistant insects.
- Many end users cannot afford it.
- Houses need to be repainted/retreated after a few years.

Mateo embedded herself in the Bolivian Chaco, the home region of indigenous peoples most impacted by Chagas disease, for extended periods. Realizing that their vulnerability to Chagas disease was directly connected to their impoverishment, she became a co-founder of the **Indigenous Women of the World Movement (MOMIM)**, which fights for social equality and cultural diversity.

She has also founded the **Pilar Mateo Foundation**, focusing on applying science and knowledge to action on behalf of those who most need it.

photo (facing page): A volunteer waters seedlings in the RisingMinds.org greenhouse, constructed from 3,000 plastic bottles and local bamboo in Ponyebar, near Lake Atitlan, Guatemala. © RisingMinds

Sector 8 – An Overview

TRANSPORTATION

Both cultural constraints and poverty hamper women's mobility. Girls and women are frequently expected to stick to the confines of their homes or local communities, though change is coming at a rapid pace as more girls go to school and more women are entering the workforce.

Impoverished girls and women unable to afford wheeled conveyances to transport themselves or their heavy burdens are forced to accomplish tasks on foot. In low-resource regions, mass transport is irregular or completely absent. Women rarely use pack animals; traditionally it is men who own, tend, and use them.

Head-carrying is a skill mastered by many low-income girls and women, particularly in Africa, but it can cause a great deal of body strain. Women are balancing loads while carrying children on their backs. And strong and hard-working as women are, they are limited to the amount of weight they can carry, consigning them to an inefficient use of their time.

Rural men are frequently migrating to cities in search of work and leaving women back in the village, giving women much more breadwinning responsibility. Women need better ways to move themselves, their children, and their goods. Their loads include water, wood for cooking fuel, produce, goods, and offspring.

Improved transportation exists: wheels. The challenge is designing women-friendly, affordable wheeled conveyances that are culturally acceptable and suitable for rugged terrain. The **Wellowater Wheel** (**#36**) for water rolling meets these criteria. **Bicycles** (**#78**) are a major upgrade, but are still too costly for most women and girls to own. All manner of handcarts, wheelbarrows, and trolleys are available, but none have displaced head-carrying as yet.

Commonplace as baby strollers are in the industrialized world, they seem to be missing from poor areas. Disabled children, though, have a promising new tool in the **Wheelchair of Hope** (**#80**), which also—literally—takes a load off their parents and allows children the independence mobility provides.

78. BIKES

Biking is much faster than walking and allows riders to transport heavier loads. Though bikes save people time and money, the poorest cannot afford them.

79. WHEELBARROWS AND HANDCARTS

Appropriately designed carts help women transport heavier loads with less physical stress.

80. AFFORDABLE CHILDREN'S WHEELCHAIRS

Lightweight, affordable wheelchairs for mobility-impaired children expand their access to education and social connection.

Biking back to school after lunch in Thiravunamali, India.
© Gina Michael

Rauha Heita, Bicycle Empowerment Network's lead bike mechanic, Okalongo, Namibia. © Michael Linke

Bikes

Biking is much faster than walking and allows riders to transport heavier loads. Though bikes save people time and money, the poorest cannot afford them.

H L&G tel

Bicycles-for-humanity.org • @WorldBicycleRelief.org • ibike.org • @IntlBike

Bicycles empower people by expanding the territory covered and the amount carried. They bring previously inaccessible opportunities into reach, helping students travel to school and raising workers' productivity.

> Bikers:
>
> Cover up to 4 times the distance a person can walk in the same time.
>
> Carry loads up to 5 times heavier than individuals.
>
> –World Bicycle Relief

Bicycles improve the lives of the poor by:

- ✔ Speeding up transportation for:
  - Rural as well as urban workers commuting to jobs
  - Farmers traveling to their fields
  - Kids biking to school, improving attendance and achievement
- ✔ Allowing for quicker load transport: water, inputs, produce, merchandise, and other people
- ✔ Allowing for quicker access to clients for professional service providers: extension services, teachers, community health workers, emergency aid
- ✔ Providing livelihood delivering merchandise **(#33)**—Eco-Fuel Africa hires teens to deliver biochar briquettes by bike, for example
- ✔ Transforming into mini-stalls to sell produce or merchandise
- ✔ Creating jobs in bike marketing and maintenance

Bikes boast additional virtues:

- ✔ Facilitating transportation that doesn't require scarce, costly, and polluting fuel
- ✔ Powering simple machines **(#30)**
- ✔ Easy sharing, multiplying the benefit of every bike

Biking to school in Siem Reap province, Cambodia. © Plan International/Mom Chantara Soleil

There is no argument about the value of bikes for alleviating poverty. The problem is their cost: $100 and up. Reused bikes donated by the affluent world hugely impact recipients, but they only meet a small fraction of the potential demand.

Buffalo Bikes, designed and manufactured by **World Bicycle Relief (WBR)**, are made for the punishing conditions of rugged Africa. The design is continually tweaked with end users, assembled in Africa from Asian-manufactured parts. WBR's bike mechanic training programs assure maintenance and the availability of replacement parts. The Buffalo is distributed through funded programs and sells for around $135.

Bikes are investments; microloans (**#88**) for bike purchases are relatively new. WBR partners with microcredit institutions to offer installment plans for bike purchases. Buyers pay off their loans through increased income realized through upfront bike ownership.

Kibbutznik and industrial designer Izhar Gafni set the Internet on fire in 2013 with videos of his cardboard bike, made out of a reported $20 of readily available recycled materials. **CardboardTech.com**, Gafni's new company, reports their bike can support 300 pounds and is waterproof, featuring wheels made of recycled car tires. A cheap, entry-level bike would be a huge poverty alleviation tool, though as of this book's writing, the bikes are not yet available.

> "Improved access to transportation, especially for the poor, can improve employment prospects, reduce the money and time spent getting to jobs and schools or hauling fuel and water, reduce the costs of inputs for small-scale enterprise activities, and increase access to markets for products."
>
> *–David Mozer, Bicycle Africa*

Balancing a heavy load uphill to De Lat, Vietnam. © Tara Alan and Tyler Kellen/ GoingSlowly.com

A bicyclist pedals her petals, Siem Reap, Cambodia. © Richard Ella

A Zambian bike recipient, Ethel takes her cousin with her to school. © Brooke Slezak/World Bicycle Relief

Bike ownership presents challenges:

- High costs are a barrier to ownership.
- Bikes require ongoing maintenance, especially when used on pitted, unpaved roads.
- Bike riding must be mastered, which is sometimes hard for adults.
- Cultural taboos prevail against women biking in some regions.
- Men control purchasing decisions and do not prioritize girls' and women's bike use or ownership.
- Women's clothing may be restrictive.
- Bicycles are easily and frequently stolen or "borrowed" and not returned.
- Safety is always an issue with bikes. Helmet use is rare.

> "Not only are students who received bicycles staying in school and passing their exams, there are other unexpected benefits, such as a reduced pregnancy rate in female students."
>
> —Follow-up Project Report, Alafia Bicycles-for-Education

Bicycle Empowerment Network Namibia, **BEN Namibia**, stresses building an infrastructure of bicycle mechanics and shops, and has trained mechanics now working in more than thirty-two communities. Bikes quickly become useless in areas with no bike repair shops; training mechanics builds a biking infrastructure, providing livelihood and keeping bikes running. BEN Namibia is a spin-off of BEN South Africa.

YOU

- **Bicycles for Humanity**, **B4H**, is a grassroots effort that after only eight years has grown to thirty-five chapters around the world. The chapters each collect 400-500 bikes and raise funds to deliver them to Sub-Saharan Africa. Each container, stocked with spare parts, becomes the nucleus of a local bicycle shop. Since 2005, they have shipped more than 75,000 bikes. But word of this successful program has quickly outstripped B4H's resources. There are 400 groups on their waiting list hoping for bikes. With more than 10 million bikes going to landfills every year, B4H is eager to create new chapters to join the effort.
- **Re-Cycle.org** does similar work in Great Britain.
- Bikers can help underwrite **World Bicycle Relief**'s distribution of their African-manufactured Buffalo bikes by joining their challenging, supported rides, or creating local or even individual bikeathons.

Fertilizer delivery at the Women's Cocoa Farming Improvement Initiative in Aceh province, Indonesia. © Robert Harmaini/ JMD Bridges to the Future

Wheelbarrows and Handcarts

Appropriately designed carts help women transport heavier loads with less physical stress.

Africart.wordpress.com

79 No simple, affordable conveyance has yet replaced African women head-loading their heavy burdens. Such a tool would need to be rugged as well as flexible, suited to ascending and descending steep, rugged terrain, traversing fields, and traveling unpaved, pitted roads, muddy in rainy seasons.

A wide variety of one- to four-wheeled carts are used around the world, including pushcarts, pull carts, wheelbarrows, and trolleys (as small rolling handcarts are called in Britain) in a myriad of styles and materials.

Old-fashioned wheelbarrows are on women farmers' wish lists because they allow them to more easily move inputs to their fields and transport their harvests, as well as equipment, to home and market. Imports from China cost around $40. Wheelbarrows are another farming implement that is affordable when shared by a women's cooperative, just like seeders and threshers **(#61).**

In remote regions, wheelbarrows can double as ambulances.

While the main motive for acquiring a bicycle (**#78**) is personal transport, bikes perform double duty as long-distance haulers. Vendors also retrofit bikes to serve as mobile mini-stalls, using them to transport their wares and then parking them and using baskets and shelves to display merchandise.

The **Wellowater** rolling water wheel, introduced in India **(#36)**, allows women to push water rather than carry it, doubling fetching capacity and hence finishing the task in half the time.

Peace Corps volunteer Arnold Wendroff served in Malawi in the late 1960s and observed how the lack of wheeled conveyances reinforces poverty. Gracefully as women move themselves and their heavy head loads—often including an infant or toddler wrapped to their backs—the practice causes a great deal of neck strain and is far less efficient than using carts. His **Malawi AfriCart** design project, a wooden cart built from plans he has posted, has become a lifelong passion.

JITA Bangladesh, a microfranchise network (**#92**), hires women to market merchandise in remote areas. "**The Last Mile Challenge**" has long bedeviled aid and development workers, health providers, and entrepreneurs. The cost of bringing goods and services to distant areas—to people with little buying power—is prohibitive. Villagers therefore miss out on opportunities. Beneficial purchases are physically and financially out of reach.

JITA supplies each micro-entrepreneur with merchandise to sell door-to-door in her village or neighborhood. Their saleswomen were constrained by the weight of carrying merchandise in heavy bags over long distances, so JITA swapped out the Santa Clause bags for **rolling trolleys**, rather like folding luggage carts, that are high enough to clear puddles and pits. This has increased entrepreneurs' carrying capacity, literally taking weight off their shoulders.

Anza Carts, lightweight, tubular, four-wheeled wagons, were marketed in Mozambique. They were shipped flat from China and assembled onsite, lowering costs.

- This field is wide open.

Affordable Children's Wheelchairs

Lightweight, affordable wheelchairs for mobility-impaired children expand their access to education and social connection.

 tel

WheelchairsofHope.org • @GoGRIT.org

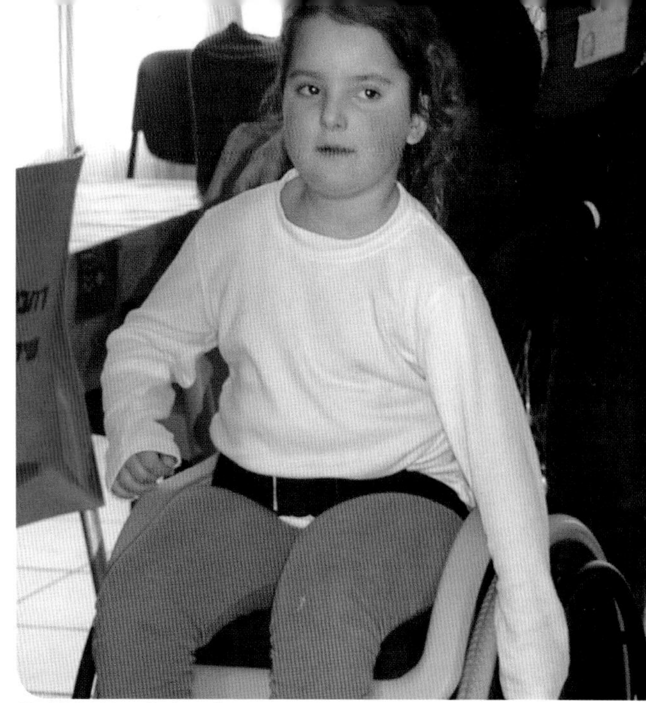

At ALYN Hospital, a girl tests out the Wheelchair of Hope, Jerusalem. © Hagai Shmueli

Extreme poverty both contributes to and exacerbates disabilities. Children born of undernourished mothers without access to maternal services are disadvantaged from birth (**#7**). Disabled children in low-resource areas have little access to treatment. Children with impaired mobility are dependent on family to transport them, a burden that falls heavily on mothers.

> "People with disabilities in developing countries are over-represented among the poorest people . . . Poverty causes disabilities and can furthermore lead to secondary disabilities for those individuals who are already disabled, as a result of the poor living conditions, health endangering employment, malnutrition, poor access to health care and education opportunities, etc. Together, poverty and disability create a vicious circle."
>
> –*The World Bank*

Wheelchair costs are prohibitive; donated, used wheelchair programs have a big impact on recipients, but—like bicycles (**#78**)—they cannot possibly fill the need.

Pablo Kaplan and Chava Rothstein, seasoned business professionals, have applied their plastics industry expertise to design **Wheelchairs of Hope** for the developing world. The pair have lifelong international business experience, but creating a product with a significant humanitarian agenda is a new direction for them. These wheelchairs:

- ✔ Weigh 10 kilograms (22 pounds), making them one-third lighter than traditional wheelchairs
- ✔ Cost about $50
- ✔ Can be assembled on-site with just twelve screws, lowering shipping costs and providing jobs
- ✔ Use bold primary colors, which make the chairs fun and less "medical" looking
- ✔ Are low maintenance and robust enough for bumpy dirt roads
- ✔ Are geared for children five to nine, but in countries where stunting is prevalent, are likely suitable for older children, too

Wheelchairs of Hope will facilitate mobility-impaired children attending school, greatly improving their future prospects for becoming self-supporting. It will also reduce their social isolation. While it is a unisex chair, disabled girls are doubly disadvantaged and stand to gain added independence from an affordable wheelchair.

> "Evidence has shown us that persons with disabilities experience disproportionately high rates of poverty and exclusion, and they lack equitable access to resources such as education, employment, and health, as well as legal and support systems. Women and girls with disabilities are disproportionately affected in all these areas."
>
> –*Lakshmi Puri, UN Women*

Wheelchairs of Hope is a social impact enterprise. The company will sell chairs directly to NGOs and intergovernmental organizations that will manage distribution systems.

Leveraged Freedom Chairs, all-terrain wheelchairs for adults in the developing world, started at the MIT Mobility Lab and has spun off as **Global Research and Technology (GRIT)**. GRIT's founders, mechanical engineers, are passionate about using technology to improve people's lives. Their design has won prestigious awards and has been through many design iterations. GRIT is also developing a model for use in the industrialized world.

YOU

- Wheelchairs of Hope is seeking Angel Investors.

Sector 9 – An Overview

INFORMATION AND COMMUNICATIONS TECHNOLOGY (ICT)

The digital revolution has changed life for most – but not all – people on the planet. Speed of communication, at ever-lowering cost, is bringing affordable digital inclusion to the world's poorest, though the digital divides remain significant. Beyond the divide of rich and poor lies another divide – one of gender. Many fewer women have access to ICT's benefits.

Information and communication technologies are increasingly hard to define: New products bridge genres and add features and components miniaturize. What is a phone versus what is a tablet may soon be a moot point. Wifi-enabled cell phones are really mini-computers. What is a radio when cell phones include them? Is it TV when episodes are viewed on a computer or a cell phone?

Countries are introduced to ICT products and systems at different rates. What is commonplace in some locales will not launch in other areas for a long time. All users of digital technology struggle with incompatible systems, wherever in the world they reside. And within an area, people adopt digital technology at varying speeds. Regions with sophisticated cellular networks often still have primitive broadband/wifi infrastructure, insufficient for transmitting large files, big data, or video streaming. In the lowest-resource regions, radios remain the most common mass communication tool.

ICT leapfrogs users. Skipping over decades of earlier models straight to cutting-edge tech makes for surprising juxtapositions. More people own cell phones than toilets. Kenyans send digital payments via mPesa all day long, whereas most Americans have not yet embraced mobile money.

ICT integrates with other sectors. Mobile payments are used for everything from bikes (**#78**) to treadle pumps (**#63**). Microinsurance coverage is a freebie added to cell phone minute purchases (**#91**). Farmers turn off irrigation equipment by cell. Women share gender-based violence incidents via info-sharing apps (**#87**). Some governments register births by digital apps. Few are untouched by ICT.

Literacy remains a barrier, along with language. Efforts are under way to translate content into less common languages; translation software for this task is still at a fairly early stage. Web users whose languages have limited presence on the web miss out on vast quantities of information.

Irregular and limited electricity constrains ICT access. Cell phone ownership is actually driving demand for rechargeable solar lanterns that also charge cell phones (**#27**). Energy-sipping computing will spread along with **distributed energy** generation (**Sector 3**); new forms of Internet connection will continue to develop to meet users' needs.

ICT costs vary enormously. While prices for cell phones and cell phone minutes are quite competitive in the Global South, other ICT costs are hard to pinpoint. Local and national ICT infrastructures vary enormously, as do national regulations; pricing is impacted by these factors, as well as by the number of providers in any given locale. The trend, though, is clear: ever more ICT access at ever lower cost.

Women and girls are underrepresented in ICT training and jobs. Worldwide, women are scarce in coding and programming and more broadly throughout the tech industry. Enhanced ICT training for girls is crucial, allowing countries to deploy girls' skills to grow their economies, benefitting all.

81. LITERACY

Reading, writing, and arithmetic mastery is perhaps the most powerful poverty alleviation tool that exists.

82. RADIO

Radio is an inexpensive medium for broadcasting local news, entertainment, and public interest content.

83. TELEVISION

Television shows portraying modern urban women with small families liberalize attitudes toward acceptable female behavior and lower birthrates.

84. CELL PHONES

Cell phones are affordable for even the very poor; their many useful functions enhance productivity—and provide entertainment.

85. COMPUTERS

Personal computers are too costly for the world's poorest billion, but affordable, frugally engineered computers are being introduced for students.

86. THE INTERNET

The Internet is a vast resource for those who can access it. Cyber cafés are one way the poor get online.

87. DIGITAL PLATFORMS FOR WOMEN'S VOICES

Digital inclusion of low-income girls and women enables them to communicate their experiences, ideas, and opinions to global audiences.

Reading in the rice paddy as Mom works nearby. © Puranjit Gangopadhyay, Courtesy of Photoshare

Literacy

Mastering reading, writing, and arithmetic is perhaps the most powerful poverty alleviation tool that exists.

Camfed.org • @camfed • RoomToRead.org • @RoomtoRead

Literacy vastly expands opportunities for the world's most economically disadvantaged. Global women's literacy rates are significantly lower than men's. While the percentage of girls in school is rising, there is still a large gender education gap, and it widens with age.

The worldwide dramas of Pakistani girls' education activist Malala Yousafzai—shot in the face by the Taliban on her school bus—and the abduction of nearly 300 Nigerian girls by Boko Haram because they were enrolled in school highlight patriarchal forces violently opposed to educating girls. Though these are radical extremists, in many regions the idea of girls' education is still new and only gradually becoming normative.

Attitudes change when educated daughters land well-paying jobs and help support their families. Educated daughters ultimately contribute more economic value for families than does farming out young daughters as indentured servants or marrying them off while still children (**#95**), a compelling economic argument for investing in girls' education.

As important as school enrollment is, it does not guarantee the acquisition of solid skills anywhere. Developing world student achievement is routinely diminished by a multitude of factors:

A serious reader, South Pacific island of Vanuatu. © Peace Corps

- ✗ Teachers who are poorly trained and/or frequently missing.
- ✗ Very large class size combined with frequent student absences.
- ✗ Attendance requiring long treks; secondary schools are typically further away than primary schools, requiring hours of daily travel or even relocation.
- ✗ Some countries do not provide public education, despite it being a Universal Human Right and universal primary education's inclusion as a Millennium Development Goal.
- ✗ Costs of uniforms, transportation, and supplies may be prohibitive, even when school is free.
- ✗ Scarcity or complete absence of textbooks, workbooks, libraries, equipment, and school supplies. Hungry students have trouble concentrating; school lunches help (**#8**).
- ✗ Undiagnosed hearing (**#14**) or vision impairment (**#15**).
- ✗ Pedagogy and curricula stressing memorization over critical thinking, failing to prepare students for creative problem-solving.

Globally, two-thirds of people over age fifteen who are illiterate are female—around 500 million. More than 50 million girls worldwide are missing from school.

All these factors affect achievement of both boys and girls, but more boys attend school, especially beyond early grades. When families' resources are constrained, boys are educated and girls stay home. Girls enrolled in school face gender-specific challenges:

- ✗ Low expectations and sexism can undermine girls' self-confidence.
- ✗ Girls' household water and wood-fetching duties, along with caring for younger siblings, hamper their school attendance and ability to keep up with class work, especially if schools are far from their homes. Girl students are at great risk for gender-based violence, both in school and during travel to and from school.
- ✗ Lack of materials to manage menstruation keeps some girls home (**#47**).
- ✗ Forced marriage curtails many schoolgirls' educations (**#95**).

Tanzanian teachers eagerly explore their new e-readers' vast resources. © Worldreader

A literacy class, Barkot, India. © Somenath Mukhopadhyay, Courtesy of Photoshare

Illiterate women have fewer earning opportunities. Children are at higher health risk when their mothers cannot read instructions or information. Illiteracy prevents people from filling out forms, accessing benefits, and navigating systems.

Record-keeping and money management are vital business skills; illiterate and innumerate women are therefore disadvantaged in commercial enterprise. Microfinance institutions (**#88**) often run literacy and numeracy programs, teaching skills to help clients become more effective entrepreneurs.

The digital revolution is rapidly expanding access to educational content and opportunities, both for those in school and older learners. Enormous local and global efforts—many highlighted in this book—are boosting literacy and numeracy.

Cell phones (**#84**) drive literacy. Mobile phones' deep penetration among the very poor has provided a tool that doubles as a handheld teacher. **Tostan** offers literacy lessons via text messages.

Worldreader.org provides loaded e-readers, vastly expanding students' accessible reading material. Refurbished e-readers are portable and rugged enough for developing world conditions. A single charge lasts a long time, so students can take them home. Partnering with **Unite For Light**, Worldreader provides backlit devices, key for students without home electricity.

In remote areas with few available teachers, radio broadcast lessons (**#82**) can provide higher-level courses, proctored by local instructors.

Computers (**#85**) and Internet access (**#86**) allow students to take advantage of learning resources such as **Khan University**, offering thousands of free video tutorials covering a vast range of subjects at all levels.

An estimated one-third of India's adult population is illiterate. **TataLiteracy.com** is a computer-based functional literacy program using teaching software, multimedia presentations, and print materials. Students learn to read within a span of thirty to forty-five hours spread over ten to twelve weeks. Indian women are more likely to be illiterate than men, and the program has been particularly successful attracting them. Many instructors are alumni of the program.

Developing world students with Internet connection can enroll in the tuition-free **University Of the People (UOP)**, enabling access to higher education without leaving home. For people with disabilities, the program has literally put learning within reach. Naylea Omayra Villanueva Sanchez, who uses a wheelchair for mobility, lives in Peru's Amazon rainforest; she was among UOP's earliest students.

Massive Online Open Courses (MOOCs), provide Web-connected students access to free, world-class university education. Pakistani schoolgirl Khadija Niazi captivated 2013 Davos Economic Forum participants when she described her experience passing **Udacity**'s über-challenging Physics 100—at age eleven.

YOU

- Become a volunteer translator for Khan Academy; their free online lessons are translated into ten languages.
- Opportunities abound to sponsor students; how about a subsidizing a teacher? **HimalayanCommunityProject.org**

YOU EDU

- **Worldreader.org** invites writers to donate written content for dissemination via digital readers.

A mother-baby group listens to a Lifeline Prime radio broadcast, Zambia. © Lifeline Energy

Radios

Radio is an inexpensive medium for broadcasting local news, entertainment, and educational programming.

H tel

LifelineEnergy.org • @LifelineEnergy
FarmRadio.org • @FarmRadio

Radios have offered people connection to the larger world for more than a century—welcoming, intimate companions providing up-to-date news, weather, sports, celebrity gossip, and music during long, dark, dimly lit evenings. Despite the expansion of TV, cell phones, and Internet, radio remains the primary communication medium for people with no electricity. Radio broadcasts are especially valuable for nonliterate audiences.

> "Because of its unrivalled access and its low production costs, radio is the technology that best meets the information and communication needs of farmers worldwide."
>
> —*FarmRadio.org*

Many basic cell phones now include radios, eliminating the need for radio batteries.

Canadian-based **FarmRadio.org** has produced content for smallholder, low-income farmers for several decades. Radio provides vital information to help farmers boost productivity, augmenting and—literally—amplifying underfunded agricultural extension services. For example, FarmRadio has teamed up with NGOs promoting the uptake of vitamin A–rich Orange-Fleshed Sweet Potato (**#59**), broadcasting a thirty-episode program dramatizing the importance of eating more nutritious food.

FarmRadio's *Her Farm Radio* meets the needs of female farmers, underserved by agricultural extension services, by:

- ✔ Choosing subjects of interest to women
- ✔ Broadcasting at times when women are available to listen
- ✔ Including women's voices on-air

FarmRadio distributes programming resources to more than 400 radio stations in thirty-eight African countries, reaching millions of farmers with locally adapted content in their native language. Talk shows target issues affecting local farming communities. FarmRadio also trains radio broadcasters.

Broadcast journalist Chouchou Namegabe, based in the Democratic Republic of the Congo, is a bold activist demanding an end to sexual violence against women; she is especially outraged by rape as a weapon of war. She has encouraged victims to speak out about their trauma on radio, breaking open taboos around the subject. This has empowered women to testify and work to bring perpetrators to justice. She founded the **South Kivu Women's Media Association**, or **AFEM** in French, an organization of female broadcasters using media to promote and defend women's rights. AFEM broadcasts to both urban and rural audiences, and sponsors women-led radio clubs.

Lifeline Energy distributes solar-powered radios and media players to some of the world's poorest communities, partnering with NGOs to supply its signature Lifeplayer MP3 ($75) and Prime ($40) radios. Designed to withstand harsh conditions, they are easily operated by illiterate users and loud enough for groups of up to sixty people to hear.

> "Radio is . . . far-reaching and can speak to communities in the most remote areas of the world. People speak of the power of the Internet, but that power is limited to a very small part of the world . . . 80 percent of the population in Sub-Saharan African use radio as their main means of communication."
>
> —*Uzma Sulaiman, Lifeline Energy*

Television

Television shows portraying modern urban women with small families liberalize attitudes toward acceptable female behavior, lowering birth rates.

PopulationMedia.org • @PopulationMedia • GlobalLeapAwards.org • @GlobalLEAPAward • PlanetRead.org

Slum dwellers rent a TV for evening watching, Kadugodi, Bangalore. © Philippe Gluck

Rich and poor alike, people enjoy passing time watching TV. Strictly speaking, televisions do not directly alleviate poverty unless they attract customers, say, to a microentrepreneur's bar or food establishment. TV watching does, however, have a powerful, well-documented impact on viewers' attitudes, bringing about behavioral changes associated with upward mobility, especially decreased family size.

Through "edutainment," people follow ongoing dramas, coming to identify with characters, adopting some of their ideas, even naming their children after them. Television exposure shapes behavior, supplementing family and peer influence.

> "Radio and television soap operas in Brazil, Ethiopia, India, Kenya, Mali, Mexico, Niger, Nigeria, Rwanda, St. Lucia, and Tanzania have been shown by independent researchers to change audience attitudes and behavior with regard to HIV/AIDS avoidance and use of family planning."
>
> —*Population Institute*

During his tenure as vice president for research at Mexico's **Televiso**, in the 1970s, Miguel Sabido pioneered the use of soap operas for social change in the developing world. The Sabido Method features characters who initially behave contrary to the value being promoted. Through plot twists and interaction with other characters, they eventually change their attitudes, coming around to more socially preferable behavior.

> "Television depresses fertility because many of its offerings provide a model of middle class families successfully grappling with the transition from tradition to modernity, helped by the fact that they have few children to support."
>
> —*Martin W. Lewis*

Using this method, the **Population Media Center** collaborates with partners in the developing world to create culturally appropriate media content. Locally important issues are dramatized in ways that people enjoy watching.

Reality TV is popular worldwide; Nigeria will soon have a novel spin on the genre, *Moma's Farm*, a series featuring male and female farmers working with draft animals pulling plows and performing other agricultural tasks. Viewers will be encouraged to vote for their favorite farmers, breaking down cultural resistance to women's use of draft animals.

PlanetRead's same-language subtitled TV shows raise literacy levels in India. When Bollywood's popular film songs are captioned, viewers can see the lyrics written and connect the text with the sound, boosting reading skills karaoke-style.

TVs that have been redesigned for off-grid viewing are hyper-efficient, running off the electricity generated by home solar systems (**#26**, **#29**). In 2014, the first **Lighting and Energy Access Partnership (LEAP)** contest for off-grid TVs took place; the competition focused public attention on this potentially immense market.

For many in the developing world, TV entertainment access will be a leapfrog tech experience. Viewers may soon be watching TV shows on handheld devices that merge cell phone and tablet, bypassing ownership of TV sets entirely.

YOU

- **Population Media** has internship positions.
- **PlanetRead** welcomes volunteers and interns in India.

Chatting in La Paz, Bolivia. © Matt Phipps

Cell Phones

Cell phones are affordable for even the very poor; their many useful functions enhance productivity—and provide entertainment.

For the world's low-income population who have never owned landline phones, cheap cell phone connectivity is **leapfrog technology**—users skip earlier stages of communication, going straight to mobile phones.

Mobile phone use among the global poor is now common and continues to expand rapidly.

Basic cell phone prices keep dropping, even as more functions are added. Battery efficiency is increasing, too, which is important for off-grid customers.

Dr. Anne Bower, a Peace Corps volunteer in Sierra Leone in the mid-1980s, recalls that phoning home required a two-day journey to Freetown, the capital city, by foot, canoe, and pickup trucks loaded with people, goats, and chickens. She then waited in line at a government office with a few phone booths. When her turn arrived, the operator connected her to America; ten-minute calls cost close to the annual income of a local farmer. Nowadays? International prepaid mobile cards cost a few cents per minute, from any location to anywhere else that has cell reception.

Cell phones integrate functions that previously were stand-alone technologies that low-income people didn't own. Cell phones provide multiple upgrades:

- ✔ Voice mail (replacing answering machines)
- ✔ Voice recording (replacing tape recorders)
- ✔ Calculator (replacing stand-alone units)
- ✔ Flashlight (supplementing or replacing battery-operated torches)
- ✔ Time and date display, timer, alarm, and stopwatch (replacing clocks and wristwatches)
- ✔ Texting (replacing telegrams and letters; indeed, Western Union no longer sends telegrams)
- ✔ Mobile money (**#90**), where available (replacing savings and checking accounts, checks, passbooks, and deposit slips)
- ✔ Radios (**#82**) are included in some models (replacing battery-operated, stand-alone units)
- ✔ Memo/notes (replacing pen and paper)
- ✔ Texts providing news, scores, prices, and weather (replacing newspapers)
- ✔ Games (supplementing playing cards, Game Boy–type devices, and small board games)

More than 250 million of the developing world's best-selling phone—the sturdy, inexpensive Nokia 1100—have been sold, making it the most popular piece of electronics ever made.

Cell phones' ability to provide entertainment and relieve boredom is of great value to people with time on their hands, rich or poor.

Watching Sesame Workshop in Kanpur, India. © Francis Gonzales, courtesy of Photoshare

Challenges:

- Typically, low-income users own phones and buy minutes separately. Their phone numbers change when they buy new SIM cards, making it hard to send health reminders or follow-up information.
- Cell phone charging is a huge hassle for off-grid users.
- Some 20 percent fewer women than men have cell phones, often because men control purchasing decisions.
- In many regions, cell phone reception is weak or entirely absent.
- Upgrading to smartphones with e-mail and Internet data plans is too costly, by far, for low-income users.

See also: **Microinsurance** (**#91**)

- Donate your old cell phone, or organize a whole phone collection drive; proceeds go to the UN GirlUp initiative: **www.GirlUp.BuyBackTech.com**

Computers

Personal computers are too costly for the world's poorest billion, but affordable, frugally engineered tablets are being introduced.

 tel

RaspberryPi.org • @Raspberry_Pi • fat-net.org • @FATtechy

A pilot group at Pardada Pardadi Educational Society use Aakash computers, Uttar Pradesh, India. © Sonali Campion

Low literacy, low income, and the lack of electricity make it difficult for the bottom billion to access computers. Stripping devices of inessential, energy-consuming features—a concept dubbed **frugal engineering**—has slashed costs, resulting in affordable, low-end computers. Though available to the general consumer market, the present focus is on computers as educational tools.

During his time at MIT, Nicholas Negroponte announced the $100 **One Laptop Per Child** initiative at the 2005 Davos World Economic Forum. Though that price goal was never met, they and other initiatives continue to work in this area.

The Indian government is committed to distributing the **Aakash** computer produced by DataWind to all of India's 220 million schoolchildren. Costing around $35, it has the potential to grow a new generation of digital natives.

The Aakash uses an Android operating system. The basic 10" wifi-enabled tablet comes pre-loaded with apps. A 7" cell phone-computer hybrid, including a camera, retails for $85. Prices will drop as demand increases and production capacity expands. Inexpensive solar chargers will be available.

Raspberry Pi is a $25 stripped-down, credit card-sized, Linux-based computer, designed to teach children computer programming and coding. Based in the United Kingdom, the Raspberry Pi Foundation has sold more than 2 million units, sharing **open-source** curricula, projects, and information for both students and teachers through their website. A $25 mouse/keyboard is also available. A Kickstarter funding project for a 9" screen exceeded its goal many times over.

UNICEF is piloting a Raspberry system for teaching Syrian students at refugee camps in Lebanon:

> "There will be basic literacy, numeracy and science, content based on Khan Academy produced by the Foundation for Learning Equality. We are also going to run a program called 'learning to code and coding to learn.' Children will be able to explore how to make games whilst also learning about their rights as a child."
>
> —*James Cranwell-Ward, UNICEF Lebanon*

Girls who grow up with computers, even if they are only available in a school lab, will master computer skills; girls who are issued their own personal devices will surely embrace digital life. But worldwide, women comprise just a small minority in computer programming and tech; efforts to narrow this gender gap are ongoing. A New Delhi-based organization, **Feminist Approach to Technology** (**FAT**), works on breaking down girls' stereotypes that techies are male: "You can't be what you can't see."

FAT has published a graphic novella written by Meenu Rawat and illustrated by emerging female artists in which girls use the power of tech to combat gender-based violence in Delhi; it's part of the **Grassroots Girls Book Club** series.

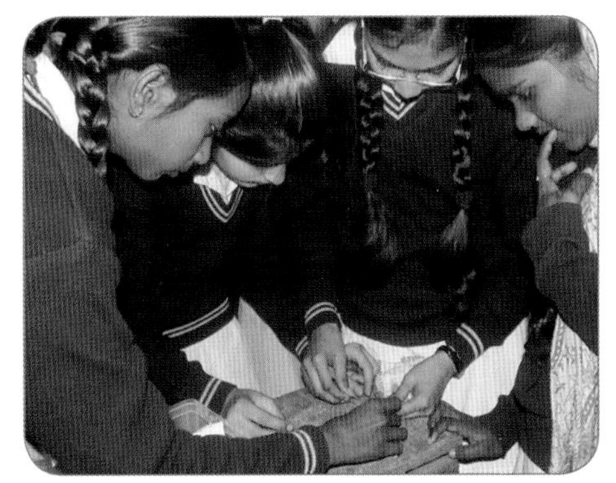

Students work on circuitry at the Senior Secondary School, Laipat Nagar, New Delhi. © Feminist Approach to Technology

YOU

- Volunteer at New Delhi's center for **Feminist Approach to Tech**.

- Download the graphic novella series at **www.GrassrootsGirls.tumblr.com**.

Surfing the 'net at the Beyond The Four Walls cyber café, Katmandu, Nepal.
© Wolf Price/ Beyond the Four Walls

The Internet

The Internet is a vast resource for those who can access it. Cyber cafés are one way the poor get online.

apc.org • @APC-News • BeyondFourWalls.org • beyond4Wwomen

86 The Internet opens up the world, providing information, entertainment, services, and communication options. However, the world's poorest billion have neither the funds to buy computers nor electricity to run them, making Internet-enabled desktops and laptops nonstarters. Their main avenues to online access are pay-per-use Internet cafés or wireless devices.

Internet access varies widely from location to location—cheap or even free in some regions, and prohibitively expensive elsewhere. Some governments—notably China, but many others, too—restrict and censure the Internet available to its citizens.

Innovative technologies are being introduced to bring Internet access to remote areas, scaling barriers of terrain, cost, and nonexistent infrastructure. Cloud services can provide powerful computing platforms at very low cost, or even for free, and minimize the need for storing data in devices.

Internet.org is a consortium committed to making the Internet—"the backbone of the knowledge economy"—available to the two-thirds of the planet not yet online. Spearheaded by Facebook's Mark Zuckerberg, the aim is to use cell phones to access the Internet. Steps to reach this goal include:

- ✔ Expanding and intensifying cell transmission infrastructure
- ✔ Reengineering phones to handle more data, using less energy
- ✔ Compressing data
- ✔ Storing phone apps internally as cache

The **Association for Progressive Communication**'s mission is introducing affordable Internet access in the developing world. The group explores ways that the Internet can narrow the digital divides of geography, income, and gender, helping low-income communities share the benefits of cyber resources. Their Women's Rights Programme is comprised of nearly 200 activists—librarians, programmers, journalists, trainers, designers, scholars, researchers, and communicators—promoting women's equality in ICT design, implementation, access, and use.

Wolf Price, a California-born **social entrepreneur**, traveled to Nepal many times, drawn by its beauty and culture. He has settled there, focusing on a mission to empower Nepali women through ICT training and providing Internet connection for isolated, impoverished girls and women. The project's name, **Beyond the Four Walls (BTFW),** contrasts Internet access with the domestic isolation of traditional Nepali women—stuck inside the four walls.

BTFW launches digital resource cyber café microenterprises run by young local women. Volunteers get free Internet; $8 buys customers a month of unlimited access. Girls and women expand their own computer skills.

> "It opens up access to crowd-funding platforms, microfinance opportunities, Skype, digital outsourcing, education, or to follow whatever creative interest women have."
>
> —*Wolf Price*

- Sales of a beautiful T-shirt featuring women rising over Nepal's Mt. Everest, modeled by a BTFW laptop user pictured above, fund Beyond The Four Walls. You can also sponsor young women's cyber café memberships directly for $8 a month.

EDU

- **Within the Four Walls**, Wolf Price's film about women and girls in Nepal, is available online.

Digital Platforms for Women's Voices

Digital inclusion of low-income girls and women enables them to communicate their experiences, ideas, and opinions to global audiences.

WorldPulse.org • @WorldPulse • InsightShare.org • @InsightShare • ushahidi.com • @Ushahidi

Documenting regional climate change in Dhanusha District, Nepal. © Pawan Kumar

NGOs have enthusiastically embraced ICT for sending women useful information, such as reminders to immunize their children or weather reports with farming advice. Less attention has been paid to ICT facilitating two-way dialogue whereby low-income women can speak up, sharing their observations and perspectives in unprecedented ways. Women now have a platform to provide first-person testimony through social media like Twitter, YouTube, Facebook, and blogging.

> "We believe technology, employed thoughtfully, will empower and amplify the voices of women and girls who have not otherwise been heard."
>
> —*SheWrites.org*

Jensine Larsen founded **World Pulse**, now a global women's communications hub engaging more than 50,000 participants from 190 countries.

World Pulse is a rich resource not only for participants but for those seeking women's first-person accounts on a wide range of issues. World Pulse provides training and mentorship, helping writers acquire the tools of the trade. Content has been picked up by global media including BBC and CNN.

World Pulse has become an international women's community. For example, Monica Islam, a World Pulse journalist from Dhaka, Bangladesh, launched a local initiative for women to map gender-based violence by location and type, like Cairo's **Harassmap**: "rape, stalking, yelling nasty propositions, beating, etc."

She asked the digital World Pulse community for advice; members from around the globe quickly pointed her to the **open-source**, free **Ushahidi Crowdmap** platform. They coached her through technical glitches, providing support and encouragement, building out the women's global digital activism network.

> "I have a burning, unquenchable hunger to know what the world will look like when women and girls experience freedom to live to their full potential."
>
> —*Jensine Larsen*

Participatory Video, engaging local people in NGO policy research initiatives, is another two-way technology. Participants are often illiterate; video bridges that gap. Grassroots input is crucial for policy decisions. With the price of video-enabled phones dropping, women will be able to capture many stories and upload them. **Insightshare.org** is an online hub for this medium.

Witness to Hunger invites low-income American women to document their experiences through photography. An initiative of Drexel University's Center for Hunger Free Communities, more than eighty-four groups have shared their images and stories, often bonding during the process. Their work has been featured by the media, publicly exhibited, and "Witnesses" were invited to the White House, along with other anti-hunger advocacy groups. When poverty has a face and a voice, people connect emotionally, more so than by just reading statistics.

See also: **Radios (#82)**

YOU

- Writers can join the **World Pulse** community.
- **World Pulse** seeks experienced coaches to virtually mentor emerging women writers and activists in the developing world.
- **World Pulse** welcomes volunteer proofreaders and translators.

EDU

- **InsightShare** runs six-day courses in participatory video in the United Kingdom.

Sector 10 – An Overview

FINANCIAL INCLUSION

Globally, poor women generally participate in informal cash economies. People living on a few dollars a day don't have a reliable cash flow. Some days there may be a bit more, but on other days much less – or zero. These fluctuations make financial management exceedingly challenging. That people survive under such difficult circumstances is a testament to human resilience and skill.

Though women have long managed household budgets, their lack of literacy and numeracy, along with traditional taboos, has mostly kept them outside the bounds of formal deal-making and money management. Financial literacy is important to overcome women's economic marginalization. Financial inclusion describes the process by which impoverished women become better able to access financial tools.

Women's financial inclusion is an important aspect of general women's empowerment. The benefits of financial inclusion apply to low-income men as well, of course. And efforts to expand financial inclusion are not limited to the developing world; many low-income residents of affluent countries are also "unbanked."

A generation ago, in Bangladesh, Muhammad Yunus spearheaded the **microcredit** movement, demonstrating that the most impoverished of women were reliable borrowers (**#88**). Small loans from an external lender, offered at fair interest rates, helped borrowers increase their earnings; some were successful enough to lift their families out of poverty.

Village Savings and Loans, managed by members, where the funds are generated and remain within the community, are a more recent development. They have been very effective, expanding across many countries (**#89**).

Women have always stashed money under their mattresses, in coffee cans, or tucked into their bodices. These savings are vulnerable to theft, eyed by family and neighbors, and subject to husbands' authority. Banks not only provide security for savings but also pay interest, allowing the money to grow. Many microfinance institutions have gradually expanded to offer a suite of microfinancial services to low-income clients: savings accounts (**#89**), insurance (**#91**), and consumer loans. Village Savings and Loans offer basic services in locations too remote for microfinance institutions.

The financial sector has gradually integrated digital technologies, in many cases eliminating the need for paper, as well as the need for travel to bank branches. This has made their services far more accessible, especially for those in remote locations. It has also lowered the costs of financial transactions. The "high touch" task of collecting microlending payments by sending agents—usually on bicycles—for face-to-face collections is giving way to cell phone payments (**#90**). Less personal, perhaps, but more efficient.

Trends in marketing can help improve life for both buyers and sellers. Microfranchises (**#92**) make it possible for goods to make it over the so-called "**Last Mile Hurdle,**" reaching customers in remote, off-grid villages. This expands consumers' options; training and support help shopkeepers hone their management skills.

Fair trade certification (**#93**) allows artisans and subsistence farmers to participate in the global economy, raising their incomes and enhancing their security. And it gives buyers the opportunity to make purchasing decisions that support global economic justice.

88. MICROLOANS

Small loans for low-income people help them start up or expand microenterprises. Microfinance institutions have gradually become bankers to the poor.

89. SAVINGS

Saving is the cornerstone of financial security. Extremely low-income people use many savings mechanisms to manage their irregular incomes.

90. MOBILE MONEY: CASH TRANSFERS BY CELL PHONE

Sending and receiving money via mobile phones enables the unbanked to increase their financial flexibility, efficiency, and security.

91. MICROINSURANCE

Affordable microinsurance policies can help financially fragile households weather the crises of illness, crop failure, or deaths and funeral expenses.

92. MICROFRANCHISES

Microfranchises bring cutting-edge supply chain management to the corner kiosk or direct to customers, providing useful products to rural communities.

93. FAIR TRADE CERTIFICATION

Fair trade links developing world farmers and artisans to global markets, improving their economic status and giving purchasers an opportunity to support them directly.

Microfinance agent recording farmers' loan repayments. © Dominic Ochola/ One Acre Fund

Microloans

Small loans for low-income people help them start up or expand microenterprises. Microfinance institutions have gradually become bankers to the poor.

tel

Milaap.org • @MilaapDotOrg •
Kiva.org • @Kiva

88 Microlending, extending credit to very poor people—usually women—was an experiment conducted by Muhammad Yunus, a Bangladeshi economist, in the 1970s. Extremely impoverished women lived near his university, and he wondered why economics couldn't come up with some practical help for these hard-working people. His answer, microcredit and the Grameen Bank, has changed the way people think about poverty, credit, and women; he won the 2006 Nobel Peace Prize for this work.

Conventional banks would not lend to impoverished people with no collateral; the illiterate and unbanked instead turned to money lenders, taking out short-term loans at very high interest rates. Though microfinance offers lower interest rates, the rates of the now 4,000-plus microfinance institutions vary greatly depending on access to capital, national regulations, and competition.

Muhammad Yunus's experiment succeeded.

- ✔ Borrowers pay back loans at a 98 percent-plus rate.
- ✔ Women invest their business profits in their families.
- ✔ Microfinance institutions provide business and financial literacy training for clients.
- ✔ Female borrowers successfully save money.

Researchers have debated for a generation whether microfinance really helps families out of poverty, without arriving at a definitive answer. There is general consensus that it helps smooth out the day-to-day instability of life at the **base of the pyramid**. The original theory, that anyone with an idea and a little capital can become a successful entrepreneur, has not proven to be true. There are successful microentrepreneurs who have grown businesses and achieved economic security; there are also borrowers with no particular entrepreneurial streak who eke out paltry livelihoods.

Original microfinance was highly regimented. Borrowers joined groups and received loans together, creating mutual support as well as peer pressure to repay. Loans could only be used for microbusinesses; personal consumer loans were not allowed. Saving (**#89**) was mandatory.

Present-day microfinance is more flexible. Mainstream banks, seeing the viability of lending to low-income clients, have entered the fray. Mobile banking (**#90**) is bringing down transaction costs. New areas for microlending include:

- ✔ Loans for investments that enhance livelihood, like bicycles (**#78**)
- ✔ Loans for housing upgrades, like latrines (**#46**)
- ✔ Integrating pay-to-own cell phone payments for solar panels (**#29**)

YOU

- Through platforms like **Kiva.org** or **Milaap.org**, people can lend money to borrowers in both the developing world and the United States. Milaap serves India and offers the opportunity to help with loans to pay school fees. The money individuals lend allows microfinance institutions to offer their clients' lowered interest rates.
- **Envia.org**, a microfinance institution in beautiful Oaxaca, Mexico, has volunteer opportunities.
- **Envia.org** also offers a weekly tour visiting clients. Tour fees fund new loans.
- **Kiva.org** offers gift cards. Recipients choose whom to lend to, receiving the proceeds when loans are repaid; those proceeds can then be lent out again.

Savings

Village Savings and Loan members do their accounting in the Huong Hoa district, Vietnam. © Pham Hong Hanh/Plan-International

Saving is the cornerstone of financial security. Extremely low-income people use many savings mechanisms to manage their irregular incomes.

 tel

vsla.net

89 Saving requires self-discipline and management skills. Researchers are impressed with how committed many low-income people are to a variety of informal savings methods. Even on erratic, meager incomes, poor people manage to accumulate capital.

Poor people function primarily in informal, cash economies; lump sums must be accumulated for large expenditures, both routine and unpredicted, including:

- Recurring expenses like rent and tuition
- Life cycle events such as births and weddings
- Seasonal holidays and festivals
- Illness, covering both costs of care and the loss of income
- Old age and funeral expenses
- Larger purchases, such as equipment, land, or other opportunities, such as bulk purchasing at a lower price, avoiding the so-called "poverty penalty"

According to Stuart Rutherford, author of *The Poor and Their Money*, the simplest savings method, beyond an individual stash, is a savings buddy. One day you entrust her with a few cents for safekeeping, another day you keep her coins, thereby lowering temptation for each and providing encouragement.

Merry-go-round savings groups typically include fifteen participants, each contributing a daily, fixed sum. The group has a set order; in fourteen days you pay numbers 1 to 14. On your day, you receive fourteen payments plus your own, a veritable windfall. Of course, participants just receive what they put in, but Day 15 feels pretty awesome.

Merry-go-rounds require physical proximity and trust, but they are an extremely motivating, effective, and fun savings method; Rutherford reports that there are thousands in existence, many that keep going for years. They build social capital and allow members to better manage their cash flow.

Microfinance institutions have included savings programs since their inception and have gradually expanded their offerings. Women with their own savings accounts have more decision-making power in how their funds will be allocated.

Village Savings and Loan Societies (VSLS) were first launched by CARE International in Niger in 1991 and have spread to sixty-plus countries. Member-owned and democratic, they are connected to larger networks that provide structure, training, and accountability.

- Groups are usually comprised of fifteen to twenty-five people.
- Members save money but also lend, splitting the proceeds from the interest paid back.
- VSLSes function as local banks in locales too remote for microfinance institutions and ignored by conventional banking.

They have proven effective at helping villagers at the lowest rung of the economic ladder manage their money and learn basic financial literacy. Member-run, they have almost no overhead expenses.

It is now possible to save by cell phone in locales where mobile money (**#90**) is available. Gone are passbooks and lockboxes; money moves electronically.

EDU

- Read Stuart Rutherford's *The Poor and Their Money*, an elegantly written look at the creative, flexible methods poor people use to help each other save money.

Filipino businesswoman Lolita Singahan uses mobile banking, eliminating the need for six-hour treks to and from her bank. © US AID

Mobile Money: Cash Transfers by Cell Phone

Sending and receiving money via mobile phones enables the unbanked to increase their financial flexibility, efficiency, and security.

S2N tel

CenterforFinancialInclusion.org • @CFI_ACCION

For the unbanked, financial transactions are extremely time-consuming, as well as insecure, being all in cash. Sending money home to family in the village requires recruiting an emissary like a friend or bus driver. The process is slow and puts both cash and courier at risk. Paying bills means physically standing in long lines. Microfinance agents have typically collected payments in person, which adds high overhead costs to borrowing.

Mobile money—sending digital cash with few keys strokes using just a basic cell phone (**#84**)—has been transformative. Mobile money offers secure cash storage while eliminating users' travel times and expenses. No more sending cash by courier or queuing up for in-person bill-paying. Integrating digital payments has simplified and lowered costs for microloan payments (**#29**, **#88**). Government aid and even paychecks can be sent directly to recipients, without significant chunks disappearing because of corruption and graft.

> "Africa is the Silicon Valley of banking. The future of banking is being defined here . . . It's going to change the world."
>
> —*Carol Realini*

Kenya's **M-PESA**—*M* is for mobile and *pesa* in Swahili means *money*—is operated by Safaricom, Kenya's largest mobile network. The company pioneered the country's mobile money industry. Quick and cheap, it has quickly caught on; in 2014, more than 68 percent of Kenyans reported using mobile money. The system is simple:

- New accounts are verified at one of more than 40,000 M-PESA outlets by trained, supervised agents.
- Each customer account is assigned a PIN. Virtual cash is stored on users' phones; the actual money sits in a bank, safe and insured. M-PESA sends a basic transaction menu by phone. The system is easily navigated, even by low-literacy clients.
- To pay bills or send money, users simply text the amount to the designated account number.
- Recipients cash out at their local M-PESA office.
- M-PESA and the agent each receive a small percentage per transaction.

Mobile money specifically benefits women by:

✔ Allowing them to send and receive payments without going far afield
✔ Lowering their risk (they're not carrying cash)
✔ Increasing their autonomy (they're managing their own personal bank accounts)

Challenges:

- Locales with multiple, incompatible networks complicate mobile money transfer.
- While mobile money has spread well beyond Kenya, countries' adoption rates differ depending on a country's banking regulations and mobile tech infrastructure.
- Mobile money systems only work within one's own country. International money transfer costs are high, presenting a major impediment for the more than 200 million global migrant workers who send remittances home to their families in their native countries.

High-income countries have been slow to adopt mobile money; customers and businesses are less motivated to adopt it, given the robust infrastructure of credit cards and checking accounts.

Microinsurance

Affordable microinsurance policies can help financially fragile households weather the crises of illness, crop failure, or deaths and funeral expenses.

 tel

MicroinsuranceNetwork.org • @NetworkFlash

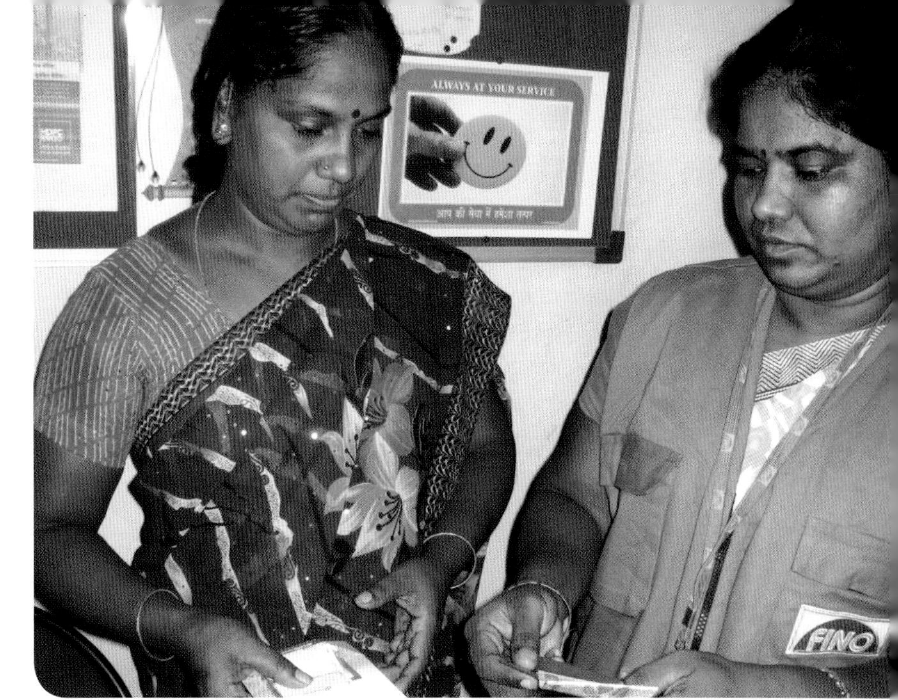

A client purchases a microinsurance policy from her insurance agent in Dharavi, a slum of Mumbai. © John Owens

Globally poor women generally participate in informal, cash-only economies with erratic income flows. Families who are stable but have no safety net are at high risk for being destabilized by catastrophes.

Muhammad Yunus launched the microfinance movement a generation ago in Bangladesh, demonstrating that even the most impoverished of women were reliable borrowers. Microloans (**#88**) typically include a savings component, as well as weekly loan repayments. Remarkably, many ultra-poor women become regular depositors.

In the 1990s, large women's groups like the **Self Employed Women's Association (SEWA)**, founded by Ela Bhatt in Ahmadabad, India, began selling insurance policies to their members. Participants' tiny premiums added up, providing a powerful path to security.

> "We're used to thinking of insurance as a safety net, but it's also a springboard."
>
> —*Andrew Kuper*

Insurance is a new product for most low-income clients. While many women manage to put aside small amounts (**#89**), a crisis with unanticipated expenses often sends a family on a downward spiral. Quick payouts help people avoid classic poverty traps like selling productive assets, pulling kids out of school, or borrowing money on unfair terms.

Insurance can help avoid the financial fallout from:

- Crop failures, when farmers have outstanding loans against their agricultural production
- The impacts of illness:
  - Expenses for treatment and medicine
  - Loss of the ill person's normal income
  - Loss of caretaker's income while caring for an ill family member
- Deaths:
  - Requiring costly funerals
  - Necessitating the settling of debts

Crop insurance, health insurance, and life insurance are financial tools that help people weather these crises, and they also provide some peace of mind to families struggling to maintain financial stability.

Microinsurance has spread quickly despite insurance being notoriously hard to market, and even more so among previously unbanked clients. When women see other women receiving cash payouts to cover health treatment and funeral expenses, they take note.

Microinsurance companies have partnered with mobile phone carriers to offer insurance coverage as a loyalty perk for cell phone minute purchasers (**#84**). The more cell phone minutes customers purchase, the more insurance they receive. Once they qualify for coverage, they can add family members to their policies and buy added coverage.

Microensure, a global microinsurance company, has teamed up with Ghana's **Tigo Mobile**; Microensure works with cellular networks in a dozen other countries as well. CEO Richard Leftly reports that they strive to pay health claims within an hour and death benefits within a day utilizing mobile payment systems (**#90**).

Crop insurance has been the most complicated to design, but it is becoming available.

Providers are also adding products tailored to women's needs, such as health insurance that includes maternity benefits. Death benefits are especially important for the newly widowed, who are extremely vulnerable in patriarchal societies (**#99**).

A JITA sales agent with customers, Rangpur (northern Bangladesh). © JITA

Microfranchises

Microfranchises bring cutting-edge supply chain management to the corner kiosk or direct to customers, providing useful products to rural communities.

LivingGoods.org • @Living_Goods • jitabangladesh.com
hapinoy.com • @Hapinoy

92 Microfinance (**#88**) aims to help women run microenterprises and earn their way up out of poverty, but not all borrowers have crackerjack business ideas or entrepreneurial talent. It's hard to come up with new retailing ideas when you've rarely been outside your small village. **Microfranchise** networks provide a vetted business model, systems, and support.

Microfranchise product mixes are continually tweaked and refined by headquarters. With solid business models to work with, many women successfully support themselves and help their villages gain access to useful consumer health products, and even small appliances, at favorable prices. Small retail shops are often better at keeping medicines in stock than underfunded government hospitals are.

There are multiple microfranchising models:

✔ Branded, stationary kiosks—usually based on groceries, but Eco-Fuel Africa has briquette kiosks; shopkeepers add whatever additional products they choose (**#33**)
✔ Door-to-door marketing—saleswomen bring merchandise direct to households, an exciting event in rural communities
✔ Portable stalls set up in the village square or regional market

Microfranchising helps shopkeepers—who are naturally risk-averse, not wanting to get stuck with inventory—expand their offerings and services. Customers benefit from having more goods available at lowered costs, due to the networks' central bulk purchasing. Being part of a larger, branded system provides livelier displays and new, useful products, helping owner's grow their store's revenue.

Distributing products to the **Last Mile** is prohibitive for social enterprises who design products that improve life for the **base of the pyramid**, like solar lanterns (**#27**), ColaLife anti-diarrhea medications (**#11**), and sanitary napkins (**#47**). Microfranchisers can bundle such products, essentially becoming distributors, and reach remote, low-income customers.

Hapinoy, a Philippine microfranchise network, supplies more than 1,000 sari-sari stores and offers microloans to its shopkeepers. Products are targeted to their low-income customer base, like selling over-the-counter medications by the tablet.

JITA Bangladesh was started by Care International to supply products to remote, under-served regions using a direct-sales Avon model. Saleswomen are called *Aparajitas*—women who never accept defeat. They roll their stocked bag directly to households (**#79**). A social business, JITA broke even two years ahead of schedule.

Living Goods, working in Uganda and Kenya, offers simple health remedies like anti-malaria treatment (**#9**). Agents are trained in basic public health screening, combining retailing with community outreach; they even make house calls.

Solar Sisters specializes in solar-powered small appliances sold in markets (**#27**). Part of Solar Sisters' mission is to educate customers about the benefits of solar.

Challenges:

- While making useful, healthful products available to purchasers, kiosks also sell cheap, high-fat, sugary junk food and drinks, contributing to worldwide obesity and tooth decay (**#6**).
- Increasing the flow of packaged products also expands the flow of trash, for which there are no waste management or recycling systems. (A good solution is bottle bricks filled with inorganic trash (**#74**).)

Fair Trade Certification

Fair trade links developing world farmers and artisans to global markets, improving their economic status and giving purchasers an opportunity to support them directly.

FairtradeUSA.org • @FairtradeUSA • WFTO.com • FairTradeJudaica.org

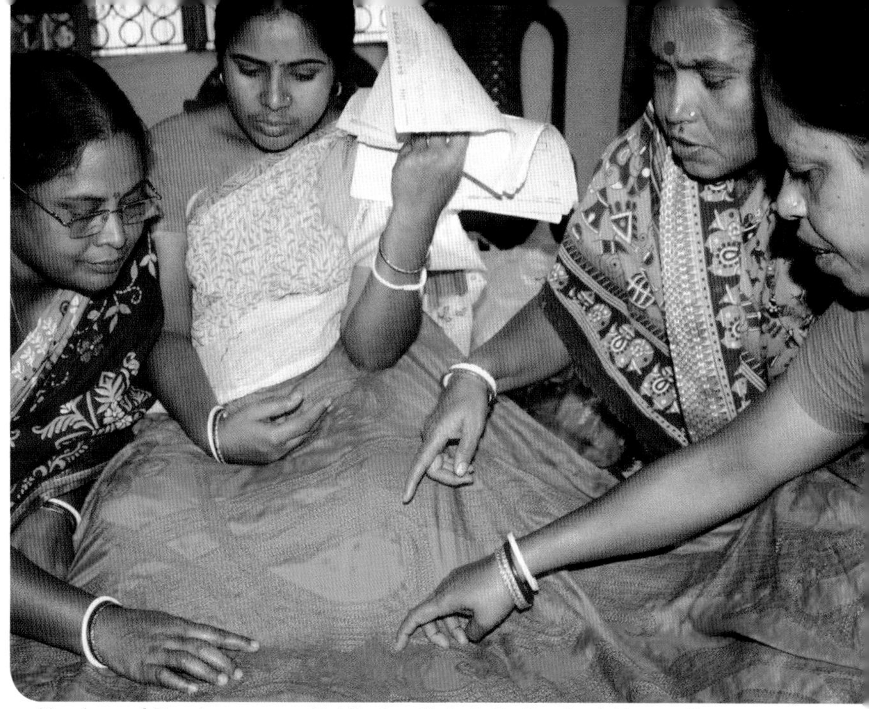

Members of Panchannagram Mahila Samiti, in Kolkotta, check the quality of a hand-embroidered bedcover ordered by Trade Aid New Zealand. © Carol Sills/WIEGO

Fair trade third-party certification tracks products through the global supply chain, assuring buyers that they are purchasing ethically sourced food and products grown by fairly compensated workers. The goal is to help low-income farmers and artisans earn more for their labor.

Fair trade partners patronize cooperative associations and work with them to educate members about meeting global market standards, incorporating environmentally friendly agricultural techniques, and promoting gender equity. Cooperatives arrange favorable loan terms for members (**#88**) and often provide health care (**#17**) and schooling for the community's children.

Fairly traded products are widely available in mainstream outlets. Customers who purchase them are willing to pay a premium for aligning their consumption with their humanitarian values. Offerings include:

✔ Coffee, tea, wine
✔ Chocolate, sugar, honey, nuts, olive oil
✔ Bananas, avocados, mangoes
✔ Flowers
✔ Apparel and accessories
✔ Crafts: ceramics, textiles, wood, metalsmith
✔ Religious ceremonial articles

Stacey Edgar founded **Global Girlfriend**, linking women's clothing and jewelry purchases to the world's neediest women to help them support themselves.

> "Since in the United States 85 percent of all brand purchases are made by women, she who holds the purse strings really can change the world. I saw Global Girlfriend as a step toward a new 'she-conomy.' "
>
> —*Stacey Edgar, Global Girlfriend Founder*

Global Girlfriend designs and tweaks artisans' products to meet the standards and tastes of the American market. They promote female-owned cooperatives. While cooperatives need to run at a profit, some offer employee support like free lunches, bikes for transportation (**#78**), and covering school fees for their employees' children (**#81**).

Challenges:

- Fair trade food and handcrafted items are pricier than their conventionally marketed competition, in part because they are higher quality. Fair-trade purchasing is generally an upscale niche, beyond many buyers' means.

Cooperative members display Agrogana Fair Trade roses, Ecuador. © Fair Trade USA

- Some fair trade customers are disheartened by how little of the cost of products winds up in the pockets of the farmers and artisans relative to retail prices.
- Fair trade certification is separate from organic certification, though products are often both organic and fair trade. This can cause confusion.
- The existence of multiple fair trade certifiers can be confusing for buyers.

- Volunteer in Ghana with **GlobalMamas**, a fair trade enterprise founded by two Peace Corps alums. "Whether you're looking for a meaningful vacation, want a career break, or are retired, we know how to put your skills to use to enrich the lives of our Mamas and their families."
- Buy products and produce that are fair trade certified.

Sector 11 – An Overview

LEGAL TOOLS

The status of women and girls is impacted by legal systems, many of which reinforce gender inequity. The Universal Declaration of Human Rights, adopted by the United Nations in 1948, enshrines female-specific rights with the weight of international law. However, nearly seventy years later, girls' and women's rights are still fragile and ignored in many parts of the world. Women's inferior legal status perpetuates poverty for everyone, male and female alike.

Women's progress out of poverty is intertwined with their legal status. While the legal tools featured here are not comprehensive, they do highlight ways that such tools can help improve girls' and women's status and decrease gender-based inequity, discrimination, and violence.

Girls and women interact with many legal systems that are sometimes at odds with each other. Tribal, ethnic, and religious legal systems impact daily life, but are not necessarily in sync with local, state, national, and international legal systems. Even when legislation protecting women's rights is on the books, laws do not guarantee social acceptance or enforcement of gender equality.

Women advocating for their rights, and their male allies, work tirelessly around the globe advancing equality for girls and women. Women's rights are human rights.

No single legal issue can be independently addressed without viewing the larger legal, social, and economic context. Any legal work that addresses the rights of the poor advances the cause of women, who are disproportionately impoverished. Working to end discrimination against minorities also benefits women; female members of minority groups who are discriminated against, such as ethnic minorities, women with disabilities, and lesbian or transgendered women, are doubly disadvantaged.

Possessing the right papers makes a big difference—a birth certificate (**#94**) and a deed to one's land (**#98**) are gateways to many benefits; their absence is poverty-perpetuating.

International treaties to combat violence against women (**#96**) in general and trafficking in specific (**#97**) give law enforcement and NGOs a framework to cooperate in addressing problems beyond the province of just one nation. They also give strength to activists within a country to pressure their government to meet international obligations.

Legal systems can improve women's well-being. Raising marriage eligibility to age eighteen, as 135 countries have done, is an important tool to prevent girls from being married off against their will (**#95**). In 128 countries, women are treated differently than men, their rights and opportunities restricted.

Rewriting legal codes to allow women titles to land (**#98**) and give them equal property inheritance rights (**#99**) is one way to expand women's security and prevent widows from falling into destitution.

Ultimately, women's votes (**#100**) and political parity will help bring gender equality to reality. Though this process is frustratingly slow, the arc of history does bend toward justice.

Relevant international laws:

CEDAW – Convention on the Elimination of all Forms of Discrimination Against Women – is the main international treaty addressing gender discrimination; 187 out of 194 countries have ratified CEDAW. The seven nonratifiers include the United States, Sudan, South Sudan, Somalia, Iran, Palau, and Tonga.

CRC – Convention on the Rights of the Child—amplifies the human rights that apply specifically to children. Unfortunately, it does not specifically mention the forced marriage of girls; the inalienable rights of many girls are denied when they are married against their will as children, including the right to education and the right to freedom from exploitation and abuse.

UDHR – Universal Declaration of Human Rights, Article 17— asserts women's land rights:

1. Everyone has the right to own property alone, as well as in association with others.
2. No one shall be arbitrarily deprived of his [sic] property.

94. UNIVERSAL BIRTH REGISTRATION

Birth registration is a gateway to education, health, and economic benefits; without birth certificates, people languish in legal limbo.

95. ERADICATING FORCED MARRIAGES OF GIRLS

Enforcing laws against forced marriages helps break the cycle of poverty they cause by curtailing girls' education.

96. INTERNATIONAL VIOLENCE AGAINST WOMEN ACT ADOPTION

Gender-based violence tramples women's rights. Violence against women is pervasive—resulting in trauma, maiming, and murder—often with impunity for perpetrators.

97. COMBATING SEX TRAFFICKING

Sex trafficking must be stopped through international enforcement and prevention; decreasing demand and increasing education are important eradication strategies.

98. LAND TITLES AND DEEDS

Land ownership provides security and an incentive to improve one's holdings, raising productivity and income. Land ownership by women confers added respect and agency.

99. INHERITANCE RIGHTS

Overturning laws barring women from inheriting their late husbands' property helps prevent widows from falling into extreme poverty.

100. THE VOTE: POLITICAL INCLUSION

Women have worked hard for political inclusion and the right to vote. Achieving legislative, executive, and judicial parity is the next goal.

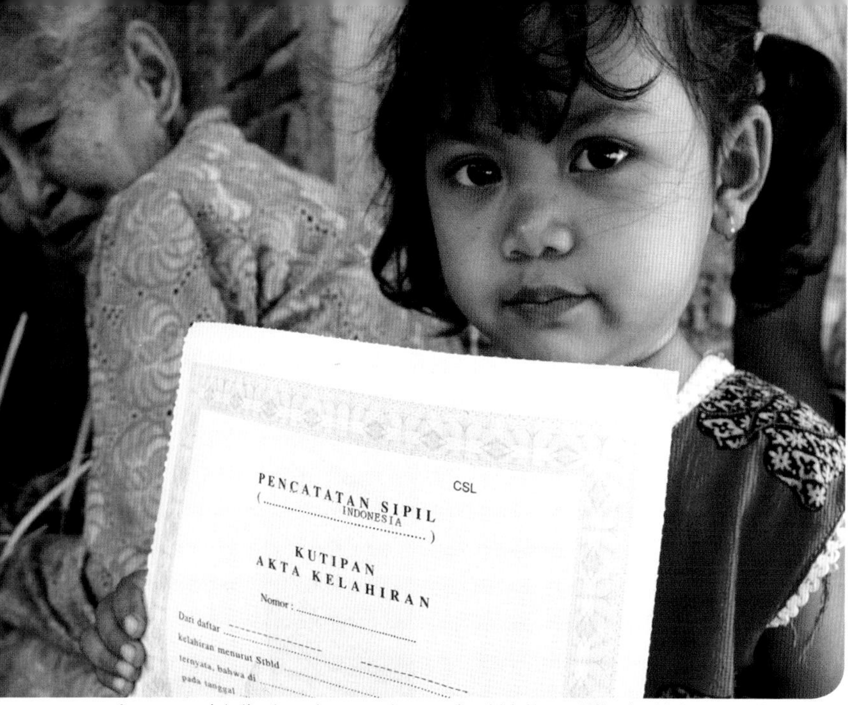

A young girl displays her newly acquired birth certificate, Kebumen, Java.
© PlanAsia/Benno Neeleman

Universal Birth Registration

Birth registration is a gateway to education, health, and economic benefits; without birth certificates, people languish in legal limbo.

Birth certificates might not seem that important to those who have them stashed in safe deposit boxes, but the absence of an official birth registration marginalizes and disadvantages individuals in a myriad of ways. Millions of births around the world are unrecorded, blocking individuals' eligibility for many government programs and benefits:

- ✗ Lack of birth documentation increases girls' vulnerability to being trafficked or married off at illegally young ages.
- ✗ Undocumented children are less likely to enroll in school or receive mandated government health services.
- ✗ Undocumented people cannot legally obtain work papers, fair legal recourse, passports, or bank accounts.
- ✗ Some countries require birth certificates for voter registration, which disenfranchises undocumented populations.
- ✗ When armed conflicts or natural catastrophes strike, displaced persons without proof of identity are more vulnerable and families are more difficult to reunite.
- ✗ Babies born in refugee camps whose births are not registered become stateless persons.
- ✗ Undocumented orphans face all these challenges alone.

Without accurate data, a country's health ministry cannot plan how to meet its children's needs. When births aren't registered, child mortality, vaccination, malnutrition, and education data go missing as well.

Article 7 of the United Nations Convention on the Rights of the Child states: ***The child shall be registered immediately after birth and shall have the right from birth to a name, [and] the right to acquire a nationality.*** **This establishes a standard for every child and holds governments accountable.**

UNICEF, The UN Refugee Agency, or UNHCR, and Plan-International have all partnered with governments to promote and implement universal birth registration, urging ministries to prioritize registration, increase transparency, and expand accessibility.

Impediments to registration include:

- ✗ Parents' failure to realize the importance of registering births.
- ✗ Bureaucratic red tape, excessive fees, and demands for bribes.
- ✗ The expense and inconvenience of traveling to far-off government offices.

Universal birth registration presents major logistical challenges, including processing the backlog of so many unrecorded children and adults. It requires reaching people in low-resource areas, who are often wary of speaking to government representatives, including:

- Nomadic peoples
- Migrant workers
- Remote indigenous tribes
- Slum families in unrecognized shantytowns
- Refugees

Systems are being improved and publicized through public awareness campaigns that emphasize the benefits of registration and guide parents through the procedures. Cambodia, for example, has raised its birth registration rate from 5 percent to 92 percent. Some additional strategies to improve these rates include:

- ✔ Digital registration, used in Kenya.
- ✔ Eliminating fees and providing remote registration.
- ✔ Tying birth registration to medical services, such as vaccinations.
- ✔ India's Unique Identification Authority assigns each citizen a twelve-digit number to simplify access to government benefits.
- ✔ In low-resource areas, The International Anglican Family Network urges churches to become birth registration sites by linking birth registration to Baptism.

Eradicating Forced Marriages of Girls

Enforcing laws against forced girl marriages helps break the cycle of poverty they cause by curtailing girls' education.

GirlsNotBrides.org • @GirlsNotBrides • ICRW.org • @ICRW

Vijaylaxmi Sharma fights to eradicate forced marriages, Rajasthan, India.
© Tanzeel Ur Rehman/Cover Asia Press, Courtesy of Photoshare

95 Marrying off girl children is illegal in most countries, but enforcement is lax or absent. Such unions are normative in many impoverished, patriarchal societies; an estimated 14 million girls are forced into marriages each year. Entering their husbands' households as *de facto* servants, they are powerless, penniless, and highly vulnerable to domestic abuse.

When daughters are considered liabilities, marrying them off means one less mouth to feed. Fathers receive a bride price. Where dowries for brides are normative, younger girls' are lower. For husbands, who are often much older, child wives provide cheap household labor. The practice is most widespread among the poorest, least-educated socioeconomic groups.

> "Marriage shall be entered into only with the free and full consent of the intending spouses."
>
> —*Universal Declaration of Human Rights Article 16:2*

Many different approaches are helping to eradicate this practice:

- ✔ In 135 countries, the legal age for marriage is already eighteen; other countries need to step up.
- ✔ Birth registration (**#94**) verifies girls' ages, helping law enforcement prevent coerced marriages.
- ✔ Human rights educations like Tostan's **Community Empowerment Program**, utilized by communities in Senegal and throughout West Africa, have inspired hundreds of communities to declare an end to child marriage and female genital mutilation (**#16**).
- ✔ Keeping girls in school, through a variety of strategies, deters forced marriages:
  - School lunches (**#8**)
  - Sanitary napkin provision (**#47**)
  - Bikes (**#78**)
  - Literacy (**#81**)
- ✔ Community organizing, led by activists like Vijaylaxmi Sharma, is effective at the village level. At age thirteen, she defied her parents, refusing marriage to a much older man. She and her parents were ridiculed. Her mother gradually accepted the wisdom of postponing marriage, impressed by reports of Sharma's academic excellence and her friend's death in childbirth. Sharma has changed attitudes one family at a time:

> "I explained that there are other options, made them understand the possibilities. Most families are illiterate, so they don't know any other life. All they need is some guidance and inspiration."
>
> —*Vijaylaxmi Sharma*

- ✔ Public health education on the danger of girls becoming pregnant before reaching full maturity raises awareness. Victims of obstetric fistula, for example, are usually teenagers. The lower the age of pregnancy, the higher the rates of maternal and infant mortality.
- ✔ 40 percent of child marriages take place in India. **Apni Beti Apna Dhan**, "Our Daughters, Our Wealth," is a conditional cash-transfer program in Haryana to incentivize valuing, investing in, and educating daughters. Unmarried eighteen-year-daughters qualify for a savings bond. Preliminary results are encouraging—the first round of potential recipients has just reached age eighteen.

EDU

- Watch *After My Garden Grows*, Megan Mylan's short film about a young girl resisting forced marriage in rural India.

An anti-rape protest in 2012, Hyderabad, India. © Swarat Ghosh

International Violence Against Women Act Adoption

Gender-based violence tramples women's rights. Violence against women is pervasive — resulting in trauma, maiming, and murder — often with impunity for perpetrators.

TheEqualityEffect.org • @Equality_Effect • FuturesWithoutViolence.org • @WithoutViolence

Girls' and women's lives are shaped by gender-based violence, circumscribing their travels, activities, and choices. Legal systems are often biased, blaming victims rather than perpetrators. Families seeking to protect daughters from sexual predators often limit girls' agency and force them into early marriages (**#95**). Vigilante violence, attacking women for perceived infractions, goes unreported and unprosecuted.

Condemnation—worldwide as well as within India—followed the 2012 gang rape of a physical therapy student. The victim ultimately died. Global protests highlighted the depravity of the behavior; within India, crowds of both women and men (one pictured above) gathered to protest women's lack of safety. The rapists were ultimately sentenced to death.

It's unclear whether violence against women has increased or women's courage to report it has increased. Either way, new attention and resources are being devoted to this scourge.

Lawyers for 160 girls aged three to seventeen successfully sued Kenya's government for failing to protect them; many of the girls were raped by men in their families. In 2013, the case was heard in the High Court, which found the police guilty of failure to enforce Kenyan laws against rape, a landmark victory for women and also for the legal systems that safeguard their rights.

The proposed **International Violence Against Women Act (IVAWA)** would be an important tool to help the United States leverage its diplomatic and development aid, promoting the criminalization and prosecution of worldwide gender-based violence and supporting intervention, education, and prevention. Drafted with input from experts from 40 international and 150 US groups, with bipartisan support, it has not (yet) passed.

IVAWA would fund local NGOs that work to combat many types of violence against girls and women, including:

- ✗ Rape as a tool of war (**#82**)
- ✗ Marital rape and sexual abuse of girls
- ✗ Forced girl marriage (**#95**)
- ✗ Murder—whether by domestic partners, honor killings, or bride burning/dowry deaths
- ✗ Acid throwing
- ✗ Female genital mutilation (**#16**)
- ✗ Sex trafficking (**#97**)
- ✗ LGBT persecution

Men are key allies in eliminating gender-based violence.

> "If we leave men out, we're essentially burdening women and girls with the task of ending this global epidemic of violence alone. Doing so also underutilizes the powerful influence that men who reject violence can have on their friends and peers."
>
> —*Brian Heilman, ICRW gender specialist*

UNite to End Violence against Women is the UN's campaign to raise public awareness and mobilize political will and resources to combat violence against women. Their goals:

- Enforcement of laws
- Coordinated action plans
- Enhanced data collection
- Increased public awareness
- Addressing sexual violence in war.

See also: **Women's Digital Activism (#87)**

YOU

- **StopStreetHarassment.org** and **TheEqualityEffect.org**, two organizations dedicated to creating a world free from gender-based violence, have a wide variety of volunteer opportunities.
- Work for the passage of **IVAWA**.

Combating Sex Trafficking

Sex trafficking must be stopped through international enforcement and prevention; decreasing demand and increasing education are important eradication strategies.

ProtectionProject.org • @ProtectionProj • FriendsofMaitiNepal.org • @FriendsofMaiti • CATWinternational.org • @CATWintl

An NGO worker checks papers at a porous border frequented by traffickers; note the camera, documenting abuses. Nepal. © Linda Egle/Sacred Threads

The sexual trafficking of girls duped into forced prostitution is a cesspool where international crime, greed, corruption, exploitation, and total disregard for human rights converge. There is little enforcement or monitoring of the kidnapping of young, uneducated, impoverished, and often undocumented girls (**#94**), in part because abductions are likely to take place across porous national borders. For brothel owners, bribes to local police are a cost of doing business.

The girls are sexually assaulted, exposed to HIV/AIDS and other STDs, abused, isolated, and threatened. Escape is a long shot; trafficked girls have no connections, money, or IDs, and they often are illiterate.

Trafficking feeds off extreme poverty, the low status of girls and women, and lax law enforcement. The International Labor Organization reports that the global commercial sex industry generates $99 billion illegal dollars per year, driven by demand: male clients pay brothels to sexually assault young coerced girls, many of whom have been lured through false promises of jobs.

One strategy to combat trafficking is to decrease demand. Male patrons are rarely apprehended or prosecuted, despite the illegality of sex with minors. **The Coalition Against Trafficking in Women (CATW)** studies and promotes enacting laws to punish purchasers of sex with children. A handful of countries are pioneers in this approach, called the **Nordic Model**.

The UN's Protocol to "**Prevent, Suppress and Punish Trafficking in Persons, Especially Women and Children**," is referred to as the **Palermo Protocol**. The United States' bill is the "**Victims of Trafficking and Violence Protection Act of 2000**." Its goal is: "To combat trafficking in persons, especially into the sex trade, slavery, and involuntary servitude, to reauthorize certain federal programs to prevent violence against women, and for other purposes."

Laws like these provide tools for law enforcement to prosecute criminal sex trafficking.

YOU

- Impoverished Nepal has a serious sex trafficking problem. Rural, uneducated girls are at high risk for abduction. **Maiti Nepal**, an NGO working to protect girls in Nepal, provides materials for trekkers to help disseminate to remote villages, via guides. If you are planning a trek, they will supply you with literature to distribute.
- **Airline Ambassadors International** trains flight attendants to become "boots in the air," watching for the telltale signs of trafficking. When their suspicions are aroused, they arrange an apprehension when the flight lands. See **AirlineAmb.org** for information about the impressive success of this program and how to receive training.
- **EternalThreads.org** markets handicrafts produced by low-income female artisans. They partner with **K.I. Nepal**, an anti-trafficking organization that runs fifteen border units – one is pictured above. Border patrols intercept up to 2,000 Nepali girls a year. **RedThreadmovement.org**, a spin-off of EternalThreads, sells bracelets made by rescued girls, raising money for their rehabilitation and promoting awareness to combat sex trafficking.

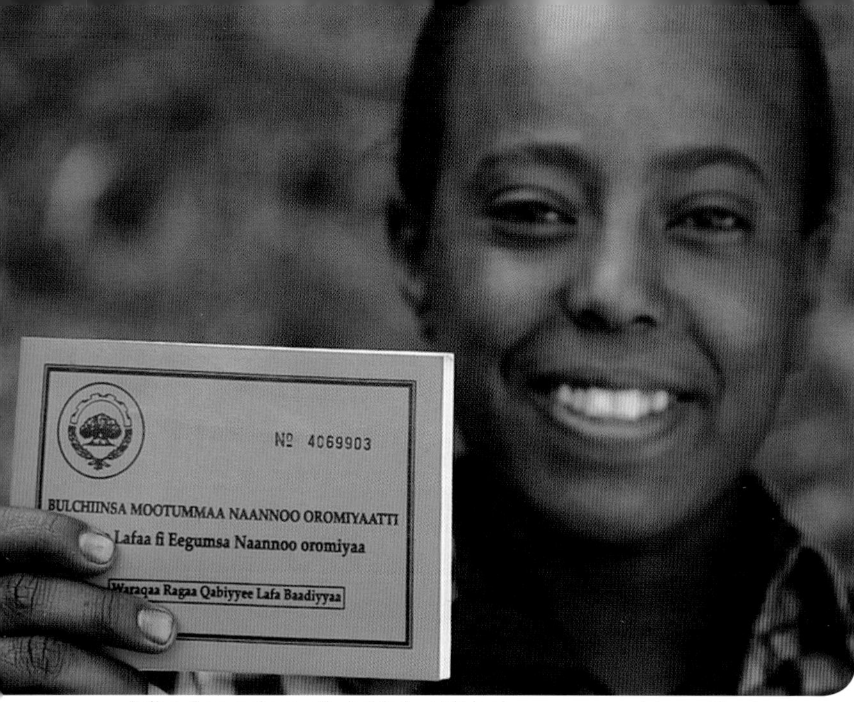

Asilya Gemmal proudly holds her Ethiopian government-issued land certificate. © Links Media/US AID Agency for International Development

Land Titles and Deeds

Land ownership provides security and an incentive to improve one's holdings, raising productivity and income. Land ownership by women confers added respect and agency.

Landesa.org • @Landesa_Global • LandCoalition.org • @LandCoalition • @WinAfrica.org • @win_Africa

98

In many regions around the world, women have never been permitted to own land. Land reform and updated laws that extend property inheritance rights to women (**#99**) are powerful poverty-alleviation tools. Informal law and customs blocking women from land ownership often conflict with formal law; educating women about their land rights and how to exercise them can help them establish more economic security.

Women's land entitlement is determined by overlapping legal systems—international, regional, national, local, religious, tribal, cultural, and ethnic—which frequently conflict. Discrimination against women within family systems contributes to their disenfranchisement. Conflict is especially intimidating to girls and women socialized to defer to men and elders. Great courage is required to stand up to male family members or land-grabbing in-laws asserting their authority.

Discriminatory laws preventing or disadvantaging women's land ownership have been successfully challenged and overthrown in Botswana, Kenya, Nigeria, South Africa, Swaziland, Tanzania, and Uganda.

> "Gender inequality hinders development: Women's insecure property rights contribute to low agricultural production, food shortages, underemployment, and poverty."
>
> —*WinAfrica.org*

Women comprise about half of the world's subsistence farmers, but own much less land than men, be it through individual or joint-ownership; rates vary greatly by country. Farmers with titles to their land are more productive; efforts to upgrade and improve their holdings are an investment in their future.

Secure land rights increase access to financial services and government services, another way in which proof of ownership enhances productivity. Activities like irrigation (**#64**), improving soil (**#57**), and tree-planting (**#69**) increase yields.

Higher farm productivity provides more food and indirectly generates more income, improving the status of farmers' children. When women's influence in family-spending decisions increases, more resources are allotted to food and education. Children of families with land titles are twice as likely to complete secondary school as those from untitled families, and they are one-third less likely to be underweight.

Adding women's names to land titles alongside their husbands' increases their role in joint decision-making. Incorporating a second voice improves decision-making and takes advantage of women's knowledge and perspectives. A study in Peru showed that when women's names were included on a family's urban land titles, there was a drop in fertility, presumably because titling increases women's bargaining power.

Domestic violence decreases when women have land titles; they command more respect, and are freer to leave violent relationships.

Regions where land ownership is unsettled experience higher degrees of conflict; this never bodes well for women.

YOU EDU

- Policy-makers, researchers, legal practitioners, and women's advocates working to strengthen women's land rights around the world – consult **Landwise.landesa.org**. Those with relevant materials are invited to add them.

Inheritance Rights

Lakshmi Venkata, a widow who inherited her husband's land, supports herself and two sons, Andhra Pradesh, India. © Deborah Espinosa

Overturning laws barring women from inheriting their late husbands' property helps prevent widows from falling into extreme poverty.

GlobalFundForWidows • @GlobalWidows • Landesa.org • @Landesa_Global • TheLoombaFoundation.org • @TheLoombaFndtn

99 Since time immemorial, widows have been emblematic of extreme poverty. In the United States, Married Women's Property Acts were only passed in the mid-19th century, allowing women to inherit late husbands' property.

In many countries in the world today, widows are still not permitted to inherit from deceased husbands, throwing women into economic chaos. Husbands' families can take back property, evicting widows from their homes and farmlands. Traumatized and lacking skills to support themselves, widows are marginalized and at high risk for abuse and exploitation; their children suffer similar fates.

Laws and customs regarding widows differ country to country. Reforming laws that discriminate against women is paramount; there are more than 100 million impoverished widows in the developing world. Laws on books don't end harassment on the ground. Education, outreach, and advocacy are essential.

The United Nations has declared June 23 International Widows Day, initiated by the **Loomba Foundation**, which focuses on improving the legal and financial status of developing world widows. Loomba and the UN are collaborating on outreach and support programs for widows in Malawi, Guatemala, and India, helping widows organize, support themselves, and defend their rights.

Legal and financial tools can redress widows' vulnerability:

- ✔ Life insurance (**#91**) pays for funeral expenses, and provides a financial cushion.
- ✔ Adding women to land titles along with their husbands (**#98**) can avoid legal limbo where male relatives jockey for land.
- ✔ Eradicating girls' forced marriages to much older men (**#95**) will decrease the number of young widows.

In many legal systems, daughters are likewise barred from inheriting their father's property. Reformers advocate for eliminating the practice of dowry, which discriminates against women in multiple ways, including severing their ability to inherit from their parents.

In a famous biblical story (Numbers 27:1-8), the daughters of Tselofehad—Mahlah, Noah, Hoglah, Milkah, and Tirzah—bring a legal claim to Moses, reasoning that since their father left no sons, his five daughters should inherit his land. (Presumably had they had a brother, the question would not have arisen.) Moses deliberates, receiving Divine counsel, and agrees that the daughters of Tselofehad are correct.

> "If a man dies and leaves no son, give his inheritance to his daughter."
>
> —*Numbers 27:8*

In contemporary Botswana, four elderly sisters sued the government for just this right—to inherit their father's land—denying a nephew's claim over theirs. Their five-year, epic legal struggle ended with a landmark 2013 victory for Edith Mmusi, age eighty, and her younger sisters Bakhani Moima, Jane Lekoko, and Mercy Ntsehkisang. Botswana's highest court, The Court of Appeal, unanimously asserted its legal power over discriminatory customary laws that denied women's inheritance rights.

See also: **Goats** (**#72**)

An elderly women casts her vote during local government elections in Rawalpindi City District, Pakistan. © Khalid Mahmood Raja, Courtesy of Photoshare

The Vote: Political Inclusion

Women have worked hard for political inclusion and the right to vote. Achieving legislative, executive, and judicial parity is the next goal.

GenderConcerns.org • @Gender_Concerns
Appropredia.org • @appropedia

100 While many take women's right to vote for granted, it is a relatively recent fact. Up until modern times, most men didn't have voting rights; that women didn't either was a nonissue. Gradual democratization in many countries expanded men's political inclusion. The women's suffrage movement culminated in the United States with the passage of the 19th Amendment, giving American women the vote in 1920. Now women are allowed to vote in most of the world's countries. Where women are excluded, men generally don't vote either.

In May 2012 local council elections, women line up for the first voting in forty-two years, Benghazi, Libya. © Megan Doherty

Women's right to vote does not always mean women participate fully, due to factors such as:

- ✗ Social conventions discouraging and restricting women's public engagement
- ✗ Limits on voter eligibility (undocumented individuals cannot register to vote and many low-income women (and men) lack birth certificates (**#94**).)
- ✗ Illiteracy making voting difficult (women's illiteracy rates are higher than men's (**#81**).)

Women's political participation is often hampered by active discrimination and inadequate resources. Though female role models and authority figures are gradually becoming less unusual, they are still the exception rather than the rule. Liberia's Ellen Johnson Sirleaf and Malawi's Joyce Banda are the second- and third-ever female heads of state in Africa. Women head governments in Latin America as well.

Female voters don't necessarily all support women's-rights agendas, which may conflict with religious ideologies. But blocs of female voters raise awareness and attention to a host of bread-and-butter concerns; nutrition, health care, maternity care, and education are often identified as "women's issues." Women are engaging more in local politics, working their way up to powerful positions and paving the way for the next wave of women, but progress is slow.

> "I strongly believe that women in leadership positions are effective brokers for peace . . ."
>
> —*Leymah Gbowee, Nobel Peace Laureate*

Women need to be at the table, working to resolve internal and international disputes. In many modern conflicts, insurgencies, and civil wars, women bear enormous burdens. They are raped, their homes are pillaged, and they are forced to flee, taking their children with them. When women have equal political representation, will our world be more peaceful? Let us hope the answer is 'Yes!'

photo (facing page): © Megan Doherty ▶

EDU

- Nobel Peace Laureate Leymah Gbowee was a leader of the Liberian women's mass movement that helped end the Second Liberian Civil War in 2003. Her story, *Pray the Devil Back to Hell*, is included—along with four others—in the PBS series *Women, War & Peace*.

AFTERWORD

More about the photographs in *100 Under $100: One Hundred Tools for Empowering Global Women*

The 150-plus photos featured in this book expand the story of women's empowerment. The photography search—finding and choosing photos, locating photographers, communicating with them around the globe to obtain permissions—was an immense and captivating project all by itself. I cannot express enough thanks to the photographers and initiatives who donated permission for their photos to be used.

Powerful images of women's transformative global work inspired me to create this book. These contrasted with the more typical portraits of women in low-resource settings: beautiful girls and women without any particular context. Such images are sometimes criticized for their exoticism, using the subjects' appearances to celebrate "the charm of the unfamiliar." Another genre I avoided was disaster photography documenting suffering, starving mothers and children in lines awaiting food and humanitarian relief. These are heartbreaking, and important for conveying information the world needs to know, but they miss the narrative of women's capabilities, strength, and resilience. That is what I have strived to communicate in *100 Under $100*.

Featured entry photos tell visual stories of women working to solve pressing problems and to improve lives; they demonstrate how these tools work, and how women utilize them. In the more abstract sections, like *Sector 11 – Legal Tools*, the photos nevertheless show women in action. The woman's strength in **#96**—protesting violence against women in India, her fist raised in outrage as she speaks to the crowd—is a formidable tool, indeed.

I am grateful for the access granted to me by two invaluable NGO photo collections.

Photoshare.org is a Public Health and Development Photography Collection. Thousands of images have been shared by humanitarian photographers to help communicate about education and global health. The collection is a service of the Knowledge for Health (K4Health) project, based at Johns Hopkins Bloomberg School of Public Health Center for Communication Programs.

© Swarat Ghosh

Photographers are invited to share their images. Development Photography Ethics are posted at their site.

PlanInternational.org was very generous in granting me access to their extensive Media Bank, sourced from the more than fifty countries where they work.

Anyone with Internet access can search what is perhaps the world's largest photo collection, Flickr.com. Many NGOs post photos there, as do individuals. It was very fortuitous that just what caught a photographer's eye was often precisely the image I was seeking, and the Internet facilitated our connection. "Yes" responses to my Flickr Mail requests made my day.

Initiatives post large numbers of photos on their Facebook pages, the source of many in the book. Between Twitter and Facebook, I was surprisingly successful in gaining permission to use casually posted photos, and for this I am very grateful and—frankly—amazed.

SUB-INDEXES

This section includes five collections of resources culled from the 100 entries, for easier access.

1. Movies and Videos
2. Women Humanitarian Tech Stars
3. Tools for Women's Cooperatives to Share
4. Travel Opportunities: Visiting Initiatives While Vacationing
5. Service Opportunities Utilizing Professional Skills

Movies and Videos

These can be screened individually, of course, but could be combined for a Women's Empowerment Film Festival.

1. ***The Cola Road***, by filmmaker Claire Ward, chronicles the pilot of ColaLife (**#11**), medication for treating life-threatening diarrhea. "In

◀ photo (facing page): Documenting regional climate change in Dhanusha District, Nepal. © Pawan Kumar

August 2012, Claire Ward spent five weeks shadowing the ColaLife team as they prepared to launch a project that would make an anti-diarrhea kit as ubiquitous as Coca-Cola in rural communities in Zambia. This is her story." E-mail claire.ward@nyu.edu to request or suggest a screening.

2. ***Solar Mamas*** is a documentary about training women solar engineers (**#26**) at India's Barefoot College. Directed by Mona Eldaief and Jehane Noujaim and produced by Mette Heide, it lasts one hour. "Learning about electrical components and soldering without being able to read, write, or understand English is the easy part. Witness Rafea's heroic efforts to pull herself and her family out of poverty."

3. Amy Smith's famous TED Talk, ***Simple Designs to Save a Life***, has been viewed nearly 700,000 times. Smith explains and demonstrates biochar briquetting (**#33**). Smith received a MacArthur "genius" grant in 2004. The talk, presented at TED2006, is 15 minutes.

4. Wolf Price, founder of a women's cyber café (**#86**) in Katmandu, Beyond the Four Walls, has been filming a documentary, ***Within the Four Walls***, about women and girls in Nepal. It is available in segments online at **www.WithintheFourWalls.com**.

5. ***After My Garden Grows*** is Megan Mylan's short feature on a girl in India opposing forced marriage (**#95**). "A young girl in rural India tills a small plot of land to feed her family and plants seeds of independence and financial freedom in her male-dominated community."

6. ***Pray the Devil Back to Hell*** is one of NPR's four-part 2011 series *Women, War, & Peace*, telling the story of women's heroic activism (**#100**) to end the Libyan civil war. "*Pray the Devil Back to Hell* is the extraordinary story of a small band of Liberian women who came together in the midst of a bloody civil war, took on the violent warlords and corrupt Charles Taylor regime, and won a long-awaited peace for their shattered country in 2003." Directed by Gini Reticker and produced by Abigail Disney in 2008; it is 72 minutes.

Women Humanitarian Tech Stars

The tech stars featured in *100 Under $100* have contributed creatively to the field of poverty alleviation tool design and dissemination. Each is profiled in the indicated entry. (There are many more women doing exciting, important work that did not fit the particular parameters of this book; I salute them all.) Additional contributions by women are included in *100 Under $100* as well.

1. (**#3**) **Jane Chen** developed the **Embrace Infant Warmer** as part of a group assignment in the Stanford University Design for Extreme Affordability course. She went on to become CEO of the resulting social enterprise devoted to developing and distributing the Embrace, engaging in extensive on-site consultation in India with end users, to produce a truly useful tool embraced by the community. Holding a master's in business administration from Stanford University and master's in public administration from Harvard University, she presently services as chief business officer of Embrace Innovations.

2. (**#28**) **Eden Full,** inventor of the **SunSaluter**, was an undergraduate at Princeton University at the time this book was written. As a recipient of the prestigious Thiel Fellowship, Full took two years off from her college studies to immerse herself in the development of her invention, which maximizes solar panels' solarization.

3. (**#30**) **Jody Wu**, founder of **Global Cycle Solutions**, focuses on harnessing pedal power to increase efficiency of domestic tasks. Trained as a mechanical engineer, she is an alumna of the MIT D-Lab and bases her social business in Tanzania.

4. (**#31**) **Katrin Puetz** studied in Germany and is earning her doctorate in agricultural engineering from the University of Hohenheim. She has started a business in Ethiopia, **BEnergy**, focusing on communally owned biodigesters. Her invention, the biogas bag, allows waste contributors to pick up and transport biogas produced by the biodigester.

5. (**#32**) **Mary Njenga** earned a doctorate in environmental science from the University of Nairobi and focuses her research on finding the optimal ingredients and ratios for eco-briquettes; she does so in an effort to help impoverished women both earn income and improve their living conditions while also improving local eco-systems.

6. (**#33**) **Amy Smith**, an MIT-trained mechanical engineer and Peace Corps alum, is a charismatic innovator and teacher, and the founder of MIT's famous, trend-setting D-Lab. She has received numerous awards and fellowships, including the prestigious MacArthur award.

7. (**#34**) **Anna Stork** and **Andrea Sreshta** met while studying at the Columbia University School of Architecture, where they shared an interest in solar tech. Their design, **LuminAID**, an inflatable LED solar light, is well-suited for disaster settings and as backup lighting for domestic power outages.

8. (**#34**) **Tricia Compas-Markman** began working on her water treatment design for disaster settings, **DayOne Response**, while studying civil engineering. She founded a chapter of Engineers without Borders at California Polytechnic State University. "On DayOne, we want to treat water in order to save lives."

9. (**#36**) **Cynthia Koenig** is the CEO of **Wello Water**, specializing in rolling water transport. She is a **social entrepreneur** with a dual MBA/MS degree from the University of Michigan's Ross School of Business and School of Natural Resources and Environment. You can watch her describe her work at TEDxGateway.

10. (**#41**) **Petra Wadström**, a Swedish microbiologist and artist, is the inventor of the **Solvatten**, a water-treatment device utilizing solar water disinfection (SODIS). She runs Solvatten as an NGO, refining and distributing Solvattens with partner NGOs. An interview with her is posted by LinkTV at YouTube, iDOBNPpVPfM.

11. (**#46**) **Sasha Kramer**, who holds a doctorate in ecology from Stanford University, is a world expert on eco-sanitation and the CEO/founder of Haiti-based **SOIL, Sustainable Organic Integrated Livelihoods**. While she did not invent eco-san, she is furthering the field significantly. Based in Haiti, a country with little sanitation for its population and which suffered a devastating earthquake in

2010, SOIL creates assets from waste. She is the lead author of an Eco-San guidebook that assists others in building eco-san systems.

12. (**#49**) **Maria Rodriguez**, founder of Guatemala-based **Byoearth**, specializes in vermiculture, utilizing worm power to transform waste into valuable assets and provide sustainable waste management. Holding a bachelor's in business administration and a master's in sustainable rural development, she shares information with other vermiculture startups around the world.

13. (**#50**) **Neha Juneja**'s entry into the crowded field of improved cookstoves, the **Grameen Greenway**, has been a runaway success. Armed with an undergraduate degree in production and industrial engineering from Delhi College of Engineering and a master's of business administration from FMS Delhi, she and co-founder Ankit Mathur consulted extensively with potential users. "Our first product, the Greenway Smart Stove, forms a modern replacement for traditional mud stoves, saving 65 percent fuel and emitting 70 percent less smoke, thus enabling healthier, happier kitchens."

14. (**#77**) **Pilar Mateo** is the inventor of **Inesfly**, an insecticide-infused paint particularly effective in preventing Chagas disease, which is transmitted by insects that breed in adobe houses. Mateo is an entrepreneur with a doctorate in chemistry, as well as a humanitarian activist.

Tools for Women's Cooperatives to Share

These allow groups to collectively afford technology that benefits all members. See **#61** on p. 80 for additional information on women's groups.

1. (**#31**) **Biodigesters** require a critical amount of biomass waste to function, but then they produce natural gas for cooking and slurry, useful for compost or fertilizer. Classic biodigesters are underground, requiring extensive excavation and construction. There are also less costly, balloon-style, aboveground biodigesters on the market. Katrin Puetz's biogas bag provides a solution for transporting noncompressed natural gas. Agricultural, livestock, and human waste are all suitable for the biodigestion process.

2. (**#30**) **Bicimáquinas**, bike-powered machines designed and fabricated by Maya Pedal, are labor-saving devices, as well as the basis for income-generating schemes that women's groups can share. They are making their plans available, helping to spread this technology.

3. (**#33**) **Eco-briquettes** can be hand-shaped in much the way women have traditionally worked dung to create fuel "cakes." However, a ratchet press—basically a mechanized briquette extruder—can be built for around $125 in materials. This speeds up the briquetting process and produces finer-quality briquettes, enhancing the viability of briquette microbusinesses. Plans are available at Legacy Foundation.

4. (**#47**) **Sanitary napkin machines**, designed by **social entrepreneur** Arunachalam Muruganantham, can be microfinanced by women's groups who pay off the loan selling the sanitary napkins they manufacture to their members and commercial customers. Muruganantham started his company, Jayaashree, after he was saw his wife using rags. He treated her to commercial napkins and realized he could design a process for manufacturing them. After tinkering for several years to produce sanitary napkins equal in quality but far cheaper than imported products, he finally succeeded.

5. (**#61**) **Plated seeders** save time and body strain when planting.

6. (**#61**) **Paddy threshers** save time and body strain when harvesting.

7. (**#68**) **Solar dehydrators** are fairly large and can process significant quantities of foods. These lend themselves to women's groups' activities, expanding usable food both for consumption and for sales.

8. (**#68**) **Universal nut shellers** are useful for processing a variety of cash crops. The Full Belly Project sells fiberglass molds to cooperatives, which then have the capacity to build four machines using local concrete. They can patronize a local welder to connect the parts. Full Belly Project has other nonelectrified agricultural machines available, too.

9. (**#73**) **Fish farming** with easily constructed cages can be done by individual women, but often a group bands together to provide security. When the cages are in public waterways, they are vulnerable to theft.

10. (**#79**) **Wheelbarrow and handcarts** are valuable aids for moving heavy supplies and crops. Sharing one (or more) can greatly increase a group's capacity.

11. (**#82**) **Radio** is an excellent medium for communication, entertainment, and education. Female farmers have far less access to training and agricultural extension services. Radios like those produced by Lifeline are well-suited for women's groups to share, gathering for informative programs and socializing.

12. (**#88**) **Microcredit** arranged through women's cooperatives or self-help groups give their members access to appropriately designed microfinance products. A cooperative can take out a loan together for a shared business enterprise; individuals can now take out loans, too, for home upgrades, school fees, and other family needs, in addition to the more traditional microenterprise loans.

13. (**#89**) **Village Savings and Loan Societies** allow women who live in remote areas beyond the reach of microfinance institutions to save and lend money, for everyone's mutual benefit.

14. (**#91**) **Microinsurance** was pioneered by a women's group, expanding microfinance services for the world's poor.

15. (**#93**) **Fair trade certification** groups work with agricultural cooperatives. As partners, they guide and advise cooperatives on improving the quality of their production to meet global standards and help provide supply chains for storing and shipping products. Access to global markets increases income. It also helps farmers and artisans plan their businesses to align their production, or designs, with world demand.

Travel Opportunities: Visiting Initiatives While Vacationing

Quite a few of this book's entries include YOU icons, indicating ways of connecting while on holiday. These actions range from the drop-off of materials to signing up for two weeks of organized volunteer activity, sometimes called **voluntourism**.

In addition to options listed here, readers planning trips abroad can search by country in the back index starting on p. 148. Initiatives of interest may welcome visitors. Arranging this in advance may allow you to bring something an organization needs, or arrange your itinerary to accommodate their programming.

Initiatives and opportunities change quickly; be sure to contact these organizations directly when planning to participate or visit.

It is important to keep expectations in check. Global poverty will not be eradicated in an hour, a day, or a week. Participants generally value their own learning as much as whatever labor they provide.

1. (**#6**) **Global Grins** combats the epidemic of developing world tooth decay by recruiting emissaries to each deliver a box of 100 toothbrushes, shipped to you upon request. I visited Oaxaca, Mexico, while writing *100 Under $100* and decided to sign on as a courier. My box of toothbrushes arrived with very simple instructions: Find a useful place to donate them. Suggestions included churches, clinics, preschools, and neighborhood centers.

   We arrived in Oaxaca late in the evening and were saddened to see so many street children peddling wares at the city square. The next morning at our beautiful B&B, a fellow guest introduced herself and explained she was on a work sabbatical, dividing her days learning Spanish and working in a mission for street children (who, it turns out, have parents). Following her lead, we went to Oaxaca Streetchildren Grassroots' center, where the toothbrushes were graciously accepted. There were no kids there at the time; happily, they were in school.

   I enjoyed the challenge of connecting with a local organization that works with impoverished children, and was pleased to know street children's needs were being addressed. It was a small but useful act.

2. (**#88**) While many people now microlend directly through **www.Kiva.org** or **www.Milaap.org**, a visit with microfinance clients offers unique immersion. On this same trip to Oaxaca, I joined **Fundación En Via**'s daylong tour to visit a half-dozen of their microfinance clients. Friends had recommended the experience; it is also—as of this writing—ranked No. 1 out of Trip Advisor's forty-two recommended activities in Oaxaca.

   We ate lunch at a new restaurant opened with a microloan and visited artisans whose profit margins had been expanded by borrowing from En Via instead of from rug resellers. We also had the pleasure of buying beautiful crafts from artisans directly (raising their profit margins even more!), though the tour is not intended to be a shopping trip. The tour guides are volunteers; our tour fees funded new loans. There are no solicitations, but while the tours don't raise money, they do raise friends for En Via. They recently crowd funded the purchase of a tour van.

3. (**#8**) **Akshaya Patra** runs an extensive network of school lunches in India, featuring state-of-the-art kitchens and systems. They welcome visitors to come and see how much they accomplish; they will be happy to involve you. Based in Bangalore, they operate in twenty locations across nine states in India. To arrange a visit, contact their American organization at **www.foodforeducation.org**.

4. (**#97**) Planning a trek in Nepal? Contact **Maiti Nepal**, an NGO working to protect girls from the active sex-trafficking trade, and they will drop off materials at your hotel. You and your guide can help disseminate the materials to remote villages where girls are at high risk.

5. (**#42**) **Potters for Peace** runs an annual winter two-week Brigade to Nicaragua that combines working with local potters, spending time in their factory, and cultural exchange. They stress that it is a rugged experience of learning and sharing, in typical developing-world conditions. Spanish is helpful, but clay is the common language.

6. (**#50**) **Stove Team International**, the designer and distributor of the Ecocina, an improved cookstove designed to meet the needs of Central American women, runs nine-day Stove Camps. Volunteers join a team tasked with building and testing stoves, and sometimes work at construction sites. Trips also include recreational tours. Spanish is useful, but not required.

7. (**#74**) **Hug It Forward** runs supported trips to Guatemala, partnering with local communities building bottle brick schools. School construction takes many months; a week's trip will share in the building process, but does not complete a school. Their weeklong trips are well-suited for families seeking hands-on service opportunities and cultural immersion.

8. (**#60**) **Semilla Nueva** runs summer trips for those interested in directly helping their farming cooperative in Guatemala. Participants are hosted by farm families, or opt for hotel accommodations farther away.

9. (**#78**) Ambitious bikers can sign up for a **World Bicycle Relief**-supported trip; proceeds help support bicycle distribution and infrastructure in Africa. Rides take place both in the United States and Africa.

10. (**#75**) The plastic thatch project of the **Reuse Everything Institute** is based in the cloud forest of Ecuador at the Maquipucuna Eco Reserve, **www.maqui.org**. This sustainable community works to preserve biodiversity and local eco-systems. Their eco-lodge is just forty minutes from Quito; proceeds help support the reserve by providing jobs for the local community and incentivizing eco-stewardship.

11. (**#31**) + (**#73**) Seeking a more open-ended adventure? Check out **www.saelaoproject.com**, a sustainable community in Nathon, Laos. SAE LAO runs an organic garden and restaurant; their biodigester produces fuel for the restaurant, fertilizer for the fields, and food for the fish ponds. Volunteers work in the fields and restaurant, teach English to local students, and perform other tasks, committing to at least four hours of daily work for a minimum of two weeks. SAE LAO posts a wish list of needed professional services on their website. They would be especially happy to host volunteers with off-grid sustainable technology and/or organic farming experience to share.

Service Opportunities Utilizing Professional Skills

It is often hard for organizations to take advantage of *pro bono* professional services. Many of the 100 entries in this book do include such opportunities. Below they are listed by professional skill; the number indicates the entry where more background can be found.

Another helpful resource for professionals looking to donate skills is **www.Catchafire.org**. Catchafire matches professionals who want to volunteer their skills with nonprofits who need their services.

COACHES:
(**#87**) **World Pulse** seeks coaches to virtually mentor emerging women writers and activists.

DENTAL PROFESSIONALS:
(**#6**) The **American Dental Association** has a portal: www.InternationalVolunteer.ada.org.

FLIGHT ATTENDANTS:
(**#97**) **Airline Ambassadors** trains volunteers to become "boots in the air," watching for sex-trafficking.

GRANT WRITERS:
(#42) Potters for Peace seeks help with grant writing.

GRAPHIC DESIGNERS:
(**#46**) **SOIL**, an eco-sanitation nonprofit in Haiti, seeks help with designing its materials.

HEALTH PRACTITIONERS:
Nurses, midwifes, doctors, and nursing, midwifery, or medical students

1. (**#17**) **Grounds for Health** actively recruits for cervical cancer screening missions to low-income communities in the developing world.
2. (**#17**) **PINCC** also actively recruits for cervical cancer screening missions to low-income communities in the developing world.
3. (**#16**) **Edna Aden University Hospital** in Somaliland welcomes volunteers.
4. (**#22**) **Life for African Mothers** welcomes skilled volunteers.
5. (**#22**) **Midwives for Haiti** welcomes volunteer midwives and relevant health professionals.

LIBRARIANS AND BOOK CATALOGUERS:
(**#16**) **Edna Aden University Hospital** in Somaliland is setting up a library and needs help.

OPTOMETRISTS:
(**#15**) **VisionSpring** offers opportunities for continuing education experience in the field.

PHOTOGRAPHERS:

1. (**#8**) **Akshaya Patra**
2. (**#30**) **Maya Pedal**
3. **Photoshare.org** is always looking for humanitarian photographers to donate images to their collection. Read more about them on p. 139.

PROOFREADERS:
(**#87**) **WorldPulse** seeks proofreaders.

RETAIL AND ONLINE MARKETING:
(**#93**) Volunteer in Ghana with **GlobalMamas**, a fair trade enterprise. "Whether you're looking for a meaningful vacation, want a career break, or are retired, we know how to put your skills to use to enrich the lives of our Mamas and their families."

STATISTICIANS AND DATA COLLECTION EXPERTS:
(**#16**) **Edna Aden University Hospital** in Somaliland welcomes volunteers to analyze data collected on female genital mutilation.

TEACHERS OF ENGLISH AS A FOREIGN LANGUAGE:
(**#16**) **Edna Aden University Hospital** in Somaliland welcomes English teachers.

TRANSLATORS:

1. (**#81**) **Khan Academy** needs translators; their lessons are offered in many languages.
2. (**#30**) **Maya Pedal** needs translators to create instruction manuals.
3. (**#87**) **World Pulse** needs translators.

VETERINARIANS:
(**#72**) **Vetswithoutborders.ca** runs trips for vets to assist in Uganda and other developing-world countries.

VIDEO EDITORS:
(**#9**) **Against Malaria** recruits volunteers to edit footage of malaria net distributions.

WRITERS:

1. (**#81**) Donate written materials to World Reader to share with students.
2. (**#30**) **Maya Pedal** needs writers for instruction manuals.

GLOSSARY

Agroforestry—Integration of trees into smallholder subsistence farming to improve productivity and income.

Alternative rites of passage—A public ceremony celebrating girls' entrance into womanhood that replaces female genital mutilation.

Appropriate technology—Technology designed with attention to the available resources at the point of use, as well as the prevailing cultural context, using the simplest level of technology to accomplish the intended task.

B Corporation–Benefit Corporation is a class of corporations providing environmental and social benefits along with benefits to stockholders.

Base of the pyramid—BOP refers to the largest, poorest global socio-economic group, the 4 billion people who live on less than $2.50 U.S. dollars a day and lack many basic infrastructures.

Biochar—The solid material obtained from the carbonization of biomass through pyrolysis (heating without the presence of oxygen.)

Biodigester—A system for processing waste anaerobically, without oxygen, to facilitate its decomposition into methane gas and nutrient-rich slurry used as fertilizer.

Black carbon—The most strongly light-absorbing component of particulate matter, it is formed by the incomplete combustion of fossil fuels, biofuels, and biomass; hence, it is major contributor to global warming. Open-fire cooking is a major source, not to mention a serious health risk to women cooking over the open flame.

Buy One Donate One—An agreement that a company will donate one item to impoverished users for each item purchased. Also called Buy One Give One.

Carbon offset funding—Companies' or individuals' subsidies of carbon reductions elsewhere in the world to offset their own carbon emissions.

Carbonization—The conversion of an organic substance into carbon or a carbon-containing residue through pyrolysis (heating without the presence of oxygen).

Cause marketing—The cooperative efforts of a for-profit business and a nonprofit organization for mutual benefit.

Co-creation—The process of engineers or designers working with end users to produce solutions that take into account cultural context, local needs, and preferences.

Declaring—The act of a community publicly declaring its commitment to end female genital mutilation.

Developing world—General term for low-income countries, often lacking sanitation and energy infrastructures.

Distributed energy—Energy generated at the site where it is consumed, in contrast to being transmitted from a central power plant. Examples include gas produced by biodigesters and electricity generated by solar panels.

Double fortified salt—Salt fortified with iodine and iron, two micronutrients in which the world's poor are frequently deficient.

Eco-san—Ecological sanitation, the recapture of the assets of excreta to produce natural gas and nutrient-rich slurry or fertilizer.

Energy poverty—The lack of access to clean, affordable fuel and electricity that forces people to rely on expensive and polluting energy sources. Examples include the direct burning of dung, wood, and crop residue for cooking, the consumption of single-use batteries for flashlights, and burning kerosene to light lanterns.

Female genital mutilation—Sometimes termed female genital cutting, it is defined by WHO as all procedures involving partial or total removal of the external female genitalia or other intentional injury to the female genital organs for nonmedical reasons. Abbreviations are FGC or FGM.

Frugal engineering—The process of reducing complexity and costs, decreasing production expenses and lowering price, to make goods more affordable for low-income purchasers. Also known as *jugaad*.

Jugaad—See "frugal engineering."

Gold standard carbon offset—A high standard of verification for carbon-reduction projects provided by Goldstandard.org.

Greywater—Water previously used for cooking, washing, or rinsing that is not contaminated by fecal content. It can be reused if safety precautions are followed.

HPV—Human papillomavirus, sexually transmitted, is the cause of most cervical cancers.

Impact investment/investors—Financial investment intended to generate a profit while also providing social and/or environmental benefits. Some angel venture capitalists invest in social businesses directly. Acumen Fund aggregates philanthropic donations and invests them in social businesses, using a hybrid nonprofit/for-profit model.

Improved cookstoves—Stoves designed to consume less fuel and save cooking time, reducing the volume of smoke produced during cooking when compared to cooking over traditional open fires; also called "clean cookstoves" and "high-efficiency cookstoves."

Last mile challenge—Distributing affordable goods to end users in remote areas without transportation infrastructures is called the "last mile challenge."

Leapfrog technology—Providing cutting-edge technology that jumps over all earlier forms of the way that service was provided. Examples include going straight to cell phones without ever having landlines, or accessing solar-powered LED lamps without ever using incandescent light bulbs.

Low-resource settings—Refers to countries or communities that are low-income and lack clean water, electricity, and sanitation infrastructures.

Market spoilers—Poorly designed products that quickly fail, sowing distrust with low-income customers and increasing their resistance to adopting subsequent versions, even when they are better designed.

MOOC—Massive Online Open Classes, free to anyone who registers, are produced by numerous groups such as Coursera, EdX, and Udacity, and transmitted via Internet. Generally they are college level, semester-long courses.

Microfinance institutions—Provider of financial services to very low income clients not serviced by conventional banks, providing loans and add-on services, such as training. Generally, but not always, nonprofit, the mission of microfinance institutions is to help their borrowers escape poverty.

Microcredit—The provision of small loans, generally at fair interest rates, to low-income borrowers whom conventional banks do not service.

NGO—Nongovernmental organization, also called a nonprofit.

Open source—Open-source technology is publicly available; designers and engineers share plans so others can access and utilize them without fees. Open-source software can be freely downloaded and used.

Pay-As-You-Go and **Pay-Till-You-Own**—Distributed solutions, generally electricity, are generated by solar panels that customers pay for at fixed intervals by cell phone. In some cases payment includes a portion intended to pay off the actual cost, so at the end of the fixed period, the customer owns the equipment. In other cases it functions as a meter, allowing end users to pay for what they consume.

Slums—Densely populated neighborhoods abutting city limits in the developing world that lack municipal infrastructures and services; also known as shantytowns, favelas, informal settlements, squatter communities, and peri-urban neighborhoods.

Permaculture—The harmonious integration of the landscape with people providing their food, energy, shelter, and other material and nonmaterial needs in a sustainable way.

Randomized controlled study—A study design that randomly assigns participants into an experimental group or a control group. As the study is conducted, the only expected difference between the control and experimental groups in a randomized controlled trial (RCT) is the outcome variable being studied.

RUTF–Ready-to-use therapeutic food—Formulated for acutely malnourished children, food packaged in tubes that children can squeeze to feed themselves, requiring no additional mixing.

Slurry—The high-nutrient, liquefied byproduct of biodigesters that can be used as fertilizer or fish food, or added to compost.

Social entrepreneur—One who merges a marketing model with achieving social benefit; some run for-profit initiatives and others are nonprofits.

SODIS—Solar disinfection is treating water by placing it in PET #1 plastic bottles and leaving it in the sun for six hours, at which point it is safe to drink.

Three Sisters—An American Indian traditional intercropping of beans, maize, and squash that increases yields and improves soil productivity.

Triple Bottom Line—For-profit companies also measuring their social and environmental impact, often summed up as "Profit, People, and Planet."

Upcycling—Reusing discarded material to create a product of a higher quality or value than the original.

Voluntourism—Traveling to a destination to provide volunteer services, informally or through an organized mission.

RESOURCES AND LINKS

An extensive listing of resources and links is posted at **www.100Under100.org**

BIBLIOGRAPHY

Alexander, Max, *Bright Lights, No City—An African Adventure on Bad Roads with a Brother and a Very Weird Business Plan*, Hyperion, 2012.

Boo, Katherine, *Beyond the Beautiful Forevers*. Random House, 2012.

Duflo, Esther, and Abhijit Banerjee, *Poor Economics*. Public Affairs, 2011.

Edgar, Stacey, *Global Girlfriends: How One Mom Made It Her Business to Help Women in Poverty Worldwide*. St. Martin's Griffin, 2012.

Gensch, R., Miso, A., Itchon, G. (2011). *Urine as Liquid Fertiliser in Agricultural Production in the Philippines—A Practical Field Guide*. Xavier University Press, Philippines, 2011 (PDF).

George, Rose, *The Big Necessity*, Henry Holt, 2008.

Gill, Kirrin, Kim Brooks, Janna McDougall, Payal Patel, Aslihan Kes, *Bridging the Gender Divide—How Technology Can Advance Women Economically.* ICRW.org, 2010 (PDF).

Haves, Emily. *Does Energy Access Help Women? Beyond Anecdotes: A Review of the Evidence*. Ashden.org, 2012 (PDF).

House, Sarah, Thérèse Mahon and Sue Cavill, *Menstrual Hygiene Matters: A Resource for Improving Menstrual Hygiene Around the World.* WaterAid.org, 2012 PDF.

IDEO.org. *Human Centered Design*, 2nd Edition. IDEO, 2011.

Jenkins, Joseph, *The Humanure Handbook—A Guide to Composting Human Manure, 3rd edition.* Joseph Jenkins Inc, 2005.

Karlan, Dean, and Jacob Appel, *More than Good Intentions*. Plume-Penguin, 2011.

Kenny, Charles, *Getting Better: Why Global Development is Succeeding—And How We Can Improve the World Even More*. Basic Books, 2011.

Kleinfeld, Rachel and Drew Sloan, *Let There Be Light: Electrifying the Developing World with Markets and Distributed Energy*. Truman National Security Institute, 2012

Kramer, Sasha, *The SOIL Guide to Eco-San*. SOIL: Sustainable Organic Integrated Livelihoods, 2012 (PDF, available on request).

Kristof, Nicholas D. and Sheryl WuDunn, *Half the Sky: Turning Oppression into Opportunity for Women Worldwide*. Knopf, 2009.

Kristof, Nicholas D. and Sheryl WuDunn, *A Path Appears: Transforming Lives, Creating Opportunity.* Knopf, 2014.

Marketing Innovative Devices for the Base of the Pyramid—A Report by Hystra. Hystra.com, 2013 (PDF).

McBrier, Page, illustrated by Lori Lohstoeter, *Beatrice's Goat*. Aladdin, 2001.

Molloy, Aimee, *However Long the Night: Molly Melching's Journey to Help Millions of African Women and Girls Triumph*. Harper One, 2013.

Nierenberg, Danielle and Brian Halweil, Project Directors, *State of the World 2011: Innovations that Nourish the Planet*. The Worldwatch Institute, 2011.

Novogratz, Jacqueline, *The Blue Sweater: Bridging the Gap Between Rich and Poor in an Interconnected World.* Rodale Books, 2010.

Olopade, Dayo, *Africa: The Bright Continent, Breaking Rules and Making Change in Modern Africa*. Houghton Mifflin Harcourt, 2014

Owen, Matthew. *Cooking Options in Refugee Situations: A Handbook of Experiences in Energy Conservation and Alternative Fuels*. UNHCR, United Nations High Commissioner for Refugees, 2002 (PDF).

Pilloton, Emily, *Design Revolution: 100 Products that Empower People*. Metropolis Books, 2009.

Polak, Paul, *Out Of Poverty: What Works When Traditional Approaches Fail.* Berrett-Koehler Publishers, 2008.

Polak, Paul, and Mal Warwick, *The Business Solution to Poverty: Designing Products and Services for Three Billion New Customers*. Berett-Koehler Publishers, 2013.

Prahalad, C.K., *Fortune at the Bottom of the Pyramid: Eradicating Poverty Through Profits*. Wharton School Publishing, 2006

Radjou, Navi, Jaideep Probhu and Simone Ahuja, *Jugaad Innovation: Think Frugal, Be Flexible, Generate Breakthrough Growth*. Jossey-Bass, 2012.

Ridley, Matt, *The Rational Optimist: How Prosperity Evolves*. Harper Perennial, 2011.

Rutherford, Stuart with Sukhwinder Arora, *The Poor and Their Money*. Practical Action, 2009.

Sharma, Ritu, *Teach a Woman to Fish: Overcoming Poverty Around the Globe*. Palgrave Macmillan Trade, 2014.

Smith, Cynthia E., *Design For The Other 90%*. Smithsonian Institution, 2007.

Surviving the First Day—State of the World's Mothers 2013. Save the Children, 2013 (PDF).

Whitfield, David and Ruth, *Cooking with Sol: Food for the Soul, Cooked with the Sun*. E-book, David-Whitfield.com.

Wilson, Nikki and Rob, *On the Up: Inspirational Stories of Social Entrepreneurs Transforming Africa.* Wripped Publications, 2012.

Wimmer, Nancy. *Green Energy for a Billion Poor: How Grameen Shakti Created a Winning Model for Social Business*. MCRV Verlag, 2012.

Yadama, Guatam N., photos by Mark Katzman, *Fires, Fuel & The Fate of 3 Billion*. Oxford University Press, 2013.

Yunus, Muhammad, with Karl Weber, *Muhammad Yunus—Creating a World Without Poverty*, Public Affairs, 2007.

photo (facing page): Tanzanian teachers Dorice Tengia (right) and Glory Mchacky eagerly explore their new e-readers' vast resources. ©Worldreader ▶

INDEX

SELECTED TITLES FROM SHE WRITES PRESS

She Writes Press is an independent publishing company founded to serve women writers everywhere.
Visit us at www.shewritespress.com.

Renewable: One Woman's Search for Simplicity, Faithfulness, and Hope by Eileen Flanagan. $16.95, 978-1-63152-968-9. At age forty-nine, Eileen Flanagan had an aching feeling that she wasn't living up to her youthful ideals or potential, so she started trying to change the world—and in doing so, she found the courage to change her life.

The Great Healthy Yard Project: Our Yards, Our Children, Our Responsibility by Diane Lewis, MD. $24.95, 978-1-938314-86-5. A comprehensive look at the ways in which we are polluting our drinking water and how it's putting our children's future at risk—and what we can do to turn things around.

Transforming Knowledge: Public Talks on Women's Studies, 1976-2011 by Jean Fox O'Barr. $19.95, 978-1-938314-48-3. A collection of essays addressing one woman's challenges faced and lessons learned on the path to reframing—and effecting—feminist change.

The Thriver's Edge: Seven Keys to Transform the Way You Live, Love, and Lead by Donna Stoneham. $16.95, 978-1-63152-980-1. A "coach in a book" from master executive coach and leadership expert Dr. Donna Stoneham, The Thriver's Edge outlines a practical road map to breaking free of the barriers keeping you from being everything you're capable of being.

Think Better. Live Better. 5 Steps to Create the Life You Deserve by Francine Huss. $16.95, 978-1-938314-66-7. With the help of this guide, readers will learn to cultivate more creative thoughts, realign their mindset, and gain a new perspective on life.

Where Have I Been All My Life? A Journey Toward Love and Wholeness by Cheryl Rice. $16.95, 978-1-63152-917-7. Rice's universally relatable story of how her mother's sudden death launched her on a journey into the deepest parts of grief—and, ultimately, toward love and wholeness.

Her Name Is Kaur: Sikh American Women Write About Love, Courage, and Faith edited by Meeta Kaur. $17.95, 978-1-938314-70-4. An eye-opening, multifaceted collection of essays by Sikh American women exploring the concept of love in the context of the modern landscape and influences that shape their lives.

Seeing Red: A Woman's Quest for Truth, Power, and the Sacred by Lone Morch. $16.95, 978-1-938314-12-4. One woman's journey over inner and outer mountains—a quest that takes her to the holy Mt. Kailas in Tibet, through a seven-year marriage, and into the arms of the fierce goddess Kali, where she discovers her powerful, feminine self.

◀ photo (facing page): © UNICEF Tanzania/ Julie Pudlowski - Preschool in Dar es Salaam